Georgia Tech
DIGITAL SIGNAL PROCESSING
LABORATORY SERIES

Speech Coding
A Computer Laboratory Textbook

Thomas P. Barnwell III
Georgia Institute of Technology and Atlanta Signal Processors, Inc.

Kambiz Nayebi
Sharif University of Technology

Craig H. Richardson
Atlanta Signal Processors, Inc.

John Wiley & Sons, Inc.
New York • Chichester • Brisbane • Toronto • Singapore

Acquisitions Editor	Steven Elliot
Marketing Manager	Debra Riegert
Production Editor	Ken Santor
Manufacturing Coordinator	Mark Cirillo

This book was typeset in 8/10 Times Roman by the authors and printed and bound by R. R. Donnelley-Harrisonburg, Inc. The cover was printed by New England Book Components, Inc.

Recognizing the importance of preserving what has been written, it is a policy of John Wiley & Sons, Inc. to have books of enduring value published in the United States printed on acid-free paper, and we exert our best efforts to that end.

System requirements for computer disk: IBM-compatible PC (80286 microprocessor or better; 80386 or better recommended); 640K RAM (4MB RAM recommended); MS-DOS; hard disk with at least 6MB of free disk space; CGA, EGA, or VGA display (VGA recommended); floating point math co-processor and standard ASCII text editor recommended. For playback and recording of signals, either a Sound Blaster compatible sound card such as the Sound Blaster Pro or Sound Blaster 16 from Creative Labs Ltd., or an Elf, Peachtree, or 320/PC-10 DSP card from Atlanta Signal Processors, Inc. is required.

Copyright ©1996, by John Wiley & Sons, Inc.

All rights reserved. Published simultaneously in Canada.

Reproduction or translation of any part of this work beyond
that permitted by Sections 107 and 108 of the 1976 United States
Copyright Act without the permission of the copyright owner is
unlawful. Requests for permission or further information should
be addressed to the Permissions Department, John Wiley & Sons, Inc.

Library of Congress Cataloging in Publication Data:
Barnwell, T. P. (Thomas Pinkney), 1943-
    Speech coding : a computer laboratory textbook / Thomas P.
Barnwell III, Kambiz Nayebi, Craig H. Richardson.
      p. cm.— (The Georgia Tech digital signal processing
laboratory series)
    Includes bibliographical references and index.
    ISBN 0-471-51692-9 (alk. paper)
    1. Speech processing systems. 2. Signal processing—Digital
techniques. I. Nayebi, Kambiz. II. Richardson, Craig H.
III. Title. IV. Series.
TK7882.S65B37 1995
621.382'8—dc20   95–31853
                                                                                 CIP

Printed in the United States of America

10 9 8 7 6 5 4 3 2 1

Foreword

After spending decades in the research laboratory, digital signal processing (DSP) is now emerging to make a significant impact on many areas of technology. As a result, DSP is becoming a basic subject in the electrical engineering curriculum. Although numerous textbooks and reference books are available to present the theory and applications of DSP, few of these books provide much in the way of "hands-on" experience that can help a student translate equations and algorithms into insight.

Experience during the past 15 years at the Georgia Institute of Technology in using computers with both basic and advanced courses in DSP has shown that the personal computer can be an extremely effective learning aid when it is combined with well-designed exercises and effective software support. The Georgia Tech Digital Signal Processing Laboratory Series builds on this teaching experience to provide a set of computer laboratory books that can be used either to supplement traditional classroom/textbook presentations of the subject or as a self-study aid.

The value of computer-based laboratory experience is clear. However, just what this experience should be is somewhat dependent on the computer resources available and on the computer skills of the students. The following three approaches have proved to be effective:

1. Provide the student with a program or set of programs that can perform specific DSP functions. In this situation, exercises are necessarily limited to running the programs on test data and observing the results.

2. Provide the student with a set of exercises that can be carried out by using a set of macros or low-level functions that can be strung together in some sort of convenient software environment. This approach has the virtue of flexibility and is much less restrictive.

3. Provide the student with test data and suggestions for projects to be car-

ried out with whatever programming resources are available. Clearly, this is the least restrictive approach, but is the most demanding of the student's programming/computer skills.

The first approach is likely to be frustratingly limited for students who are learning fundamental concepts, but it is very appropriate when the goal is to demonstrate complex algorithms that would require a great deal of time if students were to implement them on their own. For example, digital speech processing systems often combine many basic DSP functions and often have many parameters whose effects can be illustrated and studied only by using an elaborate program. Another example is filter design, where students can learn the properties of different approximation methods by simply applying those methods to the same set of specifications. At the opposite extreme is the third approach, which is obviously most suited for advanced courses or independent study where appropriate computer programming skills can be required. The second approach is perhaps the best compromise for developing insight into the fundamental algorithms and concepts of DSP. The book *Speech Coding: A Computer Laboratory Textbook* is based primarily on this approach and is the third book in the Georgia Tech Digital Signal Processing Laboratory Series. It addresses the set of topics related to the digital processing and compression of speech signals.

Speech Coding includes more than 70 exercises and projects that can be carried out using the software provided with the book. This software includes a sophisticated graphical user interface along with a wide range of basic operations and speech coding functions. At Georgia Tech, this computer laboratory mode of operation has been underway for several years. Student response to both the software and the exercises has been extremely favorable. Students appreciate the ease with which they can begin to actually do something with what they are learning in the classroom.

There is no doubt that DSP education is moving toward greater use of computers. Indeed, few subjects in the electrical engineering curriculum are so well suited to the use of computers in instruction. The Georgia Tech Digital Signal Processing Laboratory Series, whose authors have many years experience in teaching and research in the DSP field, is a valuable contribution to this emerging trend in electrical engineering education.

Ronald W. Schafer
John O. McCarty Institute Professor
Georgia Institute of Technology

Preface

This book is the result of experience gained over the past 20 years teaching undergraduate and graduate students at the Georgia Institute of Technology about speech coding, speech synthesis, and speech processing. The past several decades have been a very fruitful period for the development of new speech coding techniques and systems. This progress has been based primarily on three elements: VLSI, DSP, and commercial applications. VLSI (very large-scale integrated circuits) has given us, and continues to give us, the powerful and cost-effective digital processors necessary to make widely distributed speech coding systems economically feasible. DSP (digital signal processing) has provided most of the new techniques that have led to the rapid progress of the past 25 years. And numerous commercial applications–from telecommunications to talking toys–have provided the impetus to keep the field active.

The problem is that as a classroom or textbook topic, speech coding is potentially rather dull. Speech coding techniques are very interesting when viewed analytically or computationally, but real understanding of speech coding systems can be attained only through experimentation. As important as DSP and computational techniques have been and continue to be to speech coding, these techniques never solve the whole problem. Optimal speech coders are developed only with extensive experimentation, and speech coding techniques can be truly understood only after a great deal of experimentation. One must listen to a speech coder to understand its performance, and one must analyze it experimentally to understand why it sounds the way it does.

At Georgia Tech, we historically taught speech coding using the research computers from the Digital Signal Processing Laboratory. This often worked very well, but the approach had several drawbacks. First, the course could not "stand on its own," in the sense that the equipment required to support the course was too expensive to be justified by the course itself. Second, a great deal of effort was required to create the speech coding systems and experiments to be used with the course. This made the course laborious to teach. Third, the course laboratory materials had to be continuously updated because new computers, operating systems, and new speech coding techniques continually worked to make the laboratory materials obsolete.

This book is intended to be an effective solution to the problem of a speech coding laboratory. There is no longer any need for expensive computer systems because modern personal computers are ubiquitous and are powerful enough to support a reasonable experimental environment. In addition, sound cards such as the Sound Blaster card [1] and Atlanta Signal Processors Elf TMS320C31 and Peachtree TMS320C32 processor cards [2], supported by this laboratory software, are readily available. Likewise, personal computers now typically have the interactive graphics capabilities that are required for the time-domain and frequency-domain displays.

What this computer laboratory text represents is a combination of a book and some software. In some sense, both the book and the software can be used separately, but they are intended to be used together. The purpose of the software is to provide a highly interactive, user-friendly, self-contained speech coding workstation that allows the user to perform a wide variety of speech coding and speech processing experiments. The environment is menu driven, graphically oriented, and has a context-sensitive help capability. This software is very easy to use, and most students can use it with very little or no training.

The primary purpose of the book is to use the software to illustrate how speech coders work and sound. It does this in three ways. First, it explains the software in detail, providing a reference manual for the sophisticated user. Second, it presents and explains a set of basic speech coders analytically and in terms of the parameters that control and configure the coders. Finally, it uses exercises and projects to lead the reader through the experimental process of understanding how speech coders work and sound.

It is important to understand that this book is not intended to be used as the sole text in a course on speech coding. Rather, it is intended to be used to provide the laboratory portion of a speech course. It is, however, intended to be used by students at many different levels. The introductory undergraduate (or even high school) student can quickly learn enough to do the basic experiments and thus achieve a good overview of the basic issues in speech coding. The more sophisticated student, who can appreciate the underlying DSP techniques in depth, can achieve a truly deep understanding of the topic. Finally, the advanced student can use the software as the basis for a sophisticated speech workstation on a personal computer.

This book should be of use to anyone with an interest in speech coding systems and research. It can be used as a textbook for the laboratory portions of courses in speech coding, speech processing, speech communications, telecommunications, and DSP. It should also be of interest to engineers with design responsibility for speech communications systems, and scientists investigating speech or psychoacoustic phenomena. Finally, it should be of interest to anyone who wishes to understand at some level–introductory to advanced–exactly how speech coding systems work and sound.

In the development of this laboratory text, we were helped by many people. We would particularly like to thank Dr. Demetrius Paris, who, in his capacity as school director, provided us the initial support for this project. We would also like to give special thanks to Samuel F. Smith for developing the initial graphical user interface for the software, and for all of his support in developing the software concepts used in this book. We are also indebted to Dr. Schuyler Quackenbush, Dr. Richard Rose, Dr. Monson H. Hayes III, Dr. Roberto Bamburger, and all the reviewers who helped improve the content and presentation of this text. We also acknowledge and express our extreme appreciation to Aina Jo Barnwell for repeatedly proofreading this text. Finally, we thank our families for their unwavering love and support through the years.

Summary of the Book

Chapter 1 gives an overview of the issues in speech coding systems. This chapter explains how the properties of the human speech production system and the properties of aural perception can be used to improve the quality and reduce the bit-rate for speech coders. It also compares different classes of speech coders with regard to their data rates and use of speech properties, and it discusses how the rest of the book should be used.

The purpose of Chapter 2 is to present the experimental environment used in the remainder of the book. After discussing the installation of the software, it presents the basic concepts, and leads the reader through a tutorial on the use of the system.

Chapter 3 presents the concepts and exercises associated with signal quantization. It begins with simple fixed, uniform quantizers and continues with nonuniform, adaptive, and optimal quantizers.

Chapter 4 describes differential coding using fixed linear predictors. The speech coders covered include differential pulse code modulation (DPCM), adaptive differential pulse code modulation (ADPCM), delta modulation (DM) and adaptive delta modulation (ADM). As in all the chapters, numerous exercises and illustrations are provided.

Chapter 5 discusses a fully parametric vocoder in the form of a pitch-excited linear predictive coder (LPC). This chapter presents the basic LPC techniques, including windowing, autocorrelation analysis, covariance analysis, the Levinson recursion, gain computation, and pitch prediction. The experiments allow the user to understand how the basic parameters (window type, window length, frame interval, predictor order, etc.) in a pitch-excited LPC vocoder interact.

Chapter 6 is a synthesis of materials from Chapters 3 and 4 in the form of adaptive predictive coding (APC). The primary topics include APC without pitch prediction, APC with pitch prediction, APC with noise feedback, and the residual excited linear predictive (RELP) coder.

The subject of Chapter 7 is analysis-by-synthesis vocoders. This is one of the most recent coding techniques, and one of the most widely used and studied. The coders addressed include the multipulse-excited linear predictive coder (MPLPC), the regular pulse-excited LPC, and the code-excited linear predictive (CELP) vocoder.

Chapter 8 addresses frequency-domain coders. The emphasis in this chapter is on subband coders. The topics include two-band analysis-synthesis systems, octave-band analysis-synthesis systems, channel coders, delay compensation, and aliasing.

The last chapter is a compilation of projects for the entire book. A project is generally a large undertaking, and it is expected that only one or two would be performed as part of an average course.

Contents

1	**Introduction**	**1**
1.1	Speech Communication	2
1.2	Definition of a Speech Coder	9
1.3	Classes of Speech Coders	11
1.4	Speech Coding Standards	13
1.5	How to Use This Laboratory Text	14
2	**DSPLAB: The DSP Laboratory Software**	**17**
2.1	Introduction	17
2.2	Installing the DSPLAB Environment	17
	2.2.1 Destination Disk	18
	2.2.2 Destination Directory	18
	2.2.3 Swap Drive	18
	2.2.4 Name of Your Editor	18
	2.2.5 Type of Graphics Adapter and Monitor	19
	2.2.6 Sound Board Installation	19
	2.2.7 Modifying CONFIG.SYS and AUTOEXEC.BAT	20
	2.2.8 Verifying Installation	20
2.3	Using DSPLAB	20
	2.3.1 DSPLAB Menus and Screens	21
	2.3.2 Using the Keyboard	22
	2.3.3 Using a Mouse	24
2.4	File, Edit, and Display Menus	25
	2.4.1 File Menu	25
	2.4.2 Edit Menu	29

viii Contents

			2.4.3	Display Menu	31
			2.4.4	Freeze Display	36

3 Quantization: PCM and APCM — 41
- 3.1 Introduction — 41
- 3.2 Uniform Quantization — 43
- 3.3 Nonuniform Quantization — 46
  - 3.3.1 Logarithmic Quantizers — 49
  - 3.3.2 Optimum Quantizers — 51
- 3.4 Adaptive Quantization — 53
  - 3.4.1 Feed-forward Adaptation — 54
  - 3.4.2 Feedback Adaptation — 61
- 3.5 Exercises — 64

4 Waveform Coding with Fixed Prediction — 67
- 4.1 Introduction — 67
- 4.2 Basic DPCM — 68
- 4.3 DPCM with Adaptive Quantization (ADPCM) — 72
- 4.4 Delta Modulation (DM) — 75
  - 4.4.1 Linear Delta Modulation (LDM) — 75
  - 4.4.2 Adaptive Delta Modulation (ADM) — 78
  - 4.4.3 Continuously Variable Slope Delta Modulator (CVSD) — 80
- 4.5 Exercises — 82

5 Pitch-excited Linear Predictive Vocoder — 85
- 5.1 Introduction — 85
- 5.2 Pitch-excited LPC — 86
- 5.3 Vocal Tract Model — 88
  - 5.3.1 Correlation Computation and the LPC Analysis — 89
  - 5.3.2 Pre-emphasis — 94
  - 5.3.3 Window Considerations — 95
- 5.4 Excitation Model — 99
  - 5.4.1 Pitch Detection — 101
  - 5.4.2 Gain Computation — 101
- 5.5 Quantization of LPC Model Parameters — 102
- 5.6 Spectral Estimation Using LPC — 103
- 5.7 Exercises — 104

6 Waveform Coding with Adaptive Prediction — 107
- 6.1 Introduction — 107
- 6.2 Adaptive-Predictive Coding — 108
- 6.3 APC with Pitch Prediction (APC-PP) — 111
- 6.4 Noise-Feedback Coding — 116
- 6.5 Residual-excited LPC — 122
- 6.6 Exercises — 125

7 Analysis-by-Synthesis LPC — 127
- 7.1 Introduction 127
- 7.2 Excitation Model 129
- 7.3 Error Weighting 129
- 7.4 Analysis-by-Synthesis Procedure 131
- 7.5 Long-Term Predictors 132
- 7.6 Multipulse-excited LPC (MPLPC) 134
- 7.7 Regular Pulse-excited LPC (RPLPC) 136
- 7.8 Code-excited LPC (CELP) 137
- 7.9 Exercises 139

8 Subband Coding — 141
- 8.1 Introduction 141
- 8.2 Subband Coding 143
  - 8.2.1 Two-Band Analysis-Synthesis Systems 143
  - 8.2.2 Tree-Structured Analysis-Synthesis Systems 147
  - 8.2.3 Subband Signal Coding 150
- 8.3 Exercises 151

9 Projects — 153

Appendices — 159

A Menu Items — 159
- A.1 Menu Items by Menu Name 159
- A.2 Alphabetical List of Menu Items 165

B Extending DSPLAB — 169
- B.1 Adding a Custom D/A Driver 169
- B.2 ASPI Signal File Format 169
- B.3 Example C Program 170

C Glossary of Abbreviations — 175

Bibliography — 177

Index — 179

Introduction 1

This book is a laboratory manual for a set of experiments on speech compression systems, or, as they are more commonly known, speech coders. The goal of the book is to illustrate how the fundamental properties of the human speech production system and the human ear can be combined with digital signal processing (DSP) techniques and VLSI implementations to create modern speech coders.

In general, speech coding can be considered to be a particular specialty in the broader field of speech processing, which also includes speech analysis and speech recognition. The entire field of speech processing is currently experiencing a revolution that has been brought about by the maturation of DSP techniques and systems. Conversely, it is fair to say that speech processing has been a cradle for DSP in the sense that many of the most widely used algorithms in digital signal processing were developed or first brought into practice by people working on speech processing systems. These algorithms include digital filtering techniques, companded pulse code modulation (PCM), linear predictive coding (LPC), the short-time Fourier transform (STFT), general time-frequency representations, adaptive filtering techniques, filter bank techniques, hidden Markov models (HMM), and many more. Thus, by studying the behavior of speech processing systems, it is possible to study the behavior of many of the most important algorithms in digital signal processing.

Of all the areas within speech processing, speech coding is the most completely understood and the most mature. There are three reasons for this. The first is simply the fact that speech is one of the lowest bandwidth signals that is of truly broad interest. Telephone-quality signals generally have a bandwidth of only about 3.2 kHz, while high-quality speech can be obtained with a bandwidth of 5-6 kHz. Thus, the low sampling rates of 6400-12,000 samples per second makes real-time processing more cost effective for speech processing than for almost any other important application area.

A second significant factor in the development of speech coding systems is the continuing revolution brought by the rapid development of very large-scale integrated (VLSI) circuit technology. Of course, VLSI has had a massive impact in nearly all areas of modern technology. However, there is a special relationship between speech coding algorithms, VLSI technology, and the telecommunications industry. On the one hand, because of its low sampling rate, speech coding is usually the first DSP application area that can be addressed by a developing digital VLSI technology. On the other hand, DSP in general, and speech coding in particular, are among the few areas in which the tremendous potential of VLSI can be directly utilized in highly structured computational algorithms. Overshadowing this relationship is the tremendous economic impact of the multibillion dollar international telecommunications industry, which needs effective digital solutions in such areas as voice switching, combined voice/data systems, voice mail, mobile telephony, satellite-based mobile communications, and mobile radios. Consequently, speech coders form the basis of a uniquely well-matched set of algorithms, technologies, and applications.

The third factor of importance is the effectiveness of various DSP algorithms in solving many of the fundamental problems associated with speech coding systems. Digital signal processing techniques have proven very effective in modeling both how speech is produced and how it is perceived. This is not to say that DSP has solved all of the problems of speech coders. Nonetheless, the accomplishments in speech coding systems over the past 20 years have been phenomenal, and much of this can be attributed to the application of DSP algorithms to the speech coding problem.

1.1 Speech Communication

In its most general form, speech communications is the process of verbally transmitting an idea from one human being to another. Figure 1.1 shows a simple block diagram of the speech communications process using a speech coding system. The goal of the overall system is to correctly transmit a concept from one talker to another via the communications channel. In this process, the idea or concept is first transformed into a sentence, which is then transformed into muscular gestures of the vocal tract, throat, and lungs. The vocal tract transforms the sentence into an acoustic wave of air pressure that is received by the input microphone of the speech coding system. The task of the speech coder is to digitize the speech signal and represent it with a digital bit stream. The bit-rate of this bit stream, however, must be consistent with the transmission capacity of the channel. At the receiver, the speech decoder receives the digital bit stream and attempts to create a new speech signal that is perceptually as close to the original speech signal as possible. This signal is then transmitted acoustically into the ear of the listener, where he or she "hears" the utterance. Using both cognitive resources and an understanding of the language, the listener then "understands" the sentence and interprets its meaning.

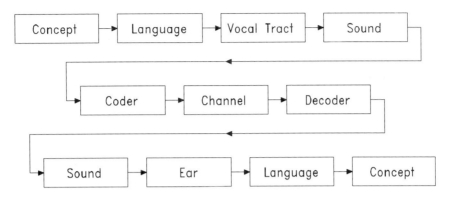

Figure 1.1. The total speech coder operating environment.

The goal of speech coder design is to produce the highest possible speech quality at the lowest possible bit-rate. The lower bound of possible bit-rates is determined by the phonemic information rate in speech, which has been estimated at about 50 bits per second [3]; and the total cognitive information rate in a speech signal, which has been estimated at about 400 bits per second. At the other extreme, the digital presentation of high-quality speech using a linear A/D system requires more than 100,000 bits per second. The challenge of a speech coder is to come as close as possible to achieving the intrinsic speech data-rate while retaining the perceived quality of the high data-rate system. If the speech coder were required to code any possible audio bandwidth signal rather than speech signals only, it would be theoretically impossible to reduce the bit-rate. The advantage that a speech coder has is that it is not required to code any possible audio bandwidth signal. Rather, *it is required to code only signals that are derived from acoustic signals produced by a human vocal tract, that represent a legal and meaningful utterance in a human language, and that will be perceived by a human listener using his or her ears.* Each of these elements–concept, language, vocal tract, and the ear–adds structure to the speech signal, and the coder can use aspects of each to remove redundancy and reduce the bit-rate. If the ideal speech coder with near-perfect quality at 400 bits per second were ever to be achieved, then it would have to take direct advantage of all of these elements. However, making use of these elements to reduce the bit-rate also has disadvantages. In general, the lower the bit-rate of a high-quality speech coder, the less robust that coder is to signals that are not "good" speech signals. Such low-rate coders often degrade dramatically for non-speech signals, for speech signals in noise, and for multiple talkers. In addition, such systems tend to be less robust when corrupted by channel bit errors and may also have long processing delays.

Cognitive Models The starting point of a speech utterance, of course, is a concept in the brain of a speaker. The brain handles and organizes concepts associatively in a manner that is not well understood. It would be ideal if the concept could in some way be communicated directly to a listener using a mechanism that does not involve speech at all. Indeed, there have been some experiments that seek to control machines directly using EEG signals. However, all work in this area is in a very primitive state (at least from an engineering perspective), and essentially no information about the cognitive process has ever been effectively used in designing a speech coding system. Thus, the use of conceptual models is an area for future research.

Language Models The use of language models in speech coding is also in a relatively primitive state, although not nearly as unrefined as the use of cognitive models. A spoken sentence has considerably more information than a written form of the same sentence. The written sentence is essentially a list of words and punctuation marks and depends greatly on the context and knowledge of the reader to remove ambiguities in its meaning. A spoken sentence has far fewer ambiguities because the speaker is aware of the intended meaning. Thus, a spoken sentence has a much more extensive representation than its written form. In addition to being just a list of words, it is also a sequence of phonemes interspersed with word and syllabic junctures. The sentence has an exact and complete syntactic structure with an associated pattern of syllable stress [4], and it also has an exact semantic representation that includes not only the meaning of the sentence, but information about the speaker's attitudes as well. This entire hierarchy of information is used by the brain in some complex fashion to create the vocal gestures that produce the speech signal.

While the use of language-related elements in speech coders is on the rise, it is not yet very common. Language models have been used successfully in very low bit-rate recognition/synthesis systems where the speech coder is similar to a speech recognition system, and the speech decoder is similar to a speech synthesis-by-rule system. Such systems have not yet been able to achieve good speech quality and are still largely a research topic. However, language models have played another important role in the design of modern speech coders. Language models provide an abstraction of the communication process that allows the significant properties of this process to be effectively described and analyzed. By studying the relationship between these abstract elements of the language and how they manifest themselves in the actual speech waveform, researchers have been able to assemble a clearer picture of the features of the acoustic waveform that should be preserved in the speech coding process.

So if cognitive models have not yet been used to advantage, and language models have played a largely secondary role, what has been the source for all the recent advances in digital speech coders? The answer lies in the development of models for the vocal tract and for aural perception.

Vocal Tract Models The human vocal tract is fundamentally a nonuniform tube whose shape varies with time. Sound is generated when an acoustic excitation signal is injected at some point in the vocal tract. For voiced sounds (such as vowels), the excitation signal is a pseudo-periodic signal generated by the vibrating vocal cords at the glottis. For unvoiced sounds (such as the letter *s*), the excitation signal is generally a strident noise source generated by turbulent air flow at a constriction in the vocal tract. Some sounds, such as voiced fricatives, can have both classes of excitations.

To a first-order approximation, the acoustic wave effects inside the vocal tract during the generation of a speech signal are linear. For this reason, it is possible to effectively model the vocal tract as a slowly time-varying acoustic filter that is excited by one or more excitation signals. More important is the fact that the vocal tract is a mechanical system. This means that the vocal tract gestures are relatively slow because they are constrained by the mass of the articulators (the tongue, jaw, lips, teeth, etc.). This results in a vocal tract model where the linear filter varies relatively slowly with time. If the vocal tract filter can be realized using a parametric filter model, the parameters of that filter will vary slowly with time. Therefore, the underlying bit-rates associated with the vocal tract filter parameters will be lower than the speech signal itself. This is often called the *short-time stationarity* property of the vocal tract filter.

The excitation signal also has properties that can be represented parametrically. During voiced sounds, the primary excitation for the vocal tract filter comes from vibrations of the glottis. The frequency of vibration at the glottis is called the *fundamental frequency*, and is perceived as the pitch of the voiced sound. The pitch pattern of a speech signal is easily perceived by listeners and carries a great deal of information. However, it is the pitch itself–that is, the rate of vibration–and not the details of the variation that are perceptually important. Due to the physiology of the glottis, the fundamental frequency of vibration does not vary greatly with time. Thus, the voiced excitation can often be modeled adequately using a slowly varying model of the fundamental frequency. This *pitch signal* can be represented with a much lower bit-rate than the voiced excitation signal itself.

The two parametric speech models that have been used the most in modern speech coders are the *pitch-excited vocoder model* shown in Figure 1.2 and the *waveform-excited vocoder model* shown in Figure 1.3. Both models present the vocal tract filter as a slowly varying, parametrically controlled linear filter. By far, the most successful and widely used parametric filter is an all-pole fully recursive filter whose parameters can be derived using linear predictive analysis. This class of coders is often called *linear predictive coders* (LPCs). However, a number of other types of time-varying filters have been used to advantage, including filter bank structures (channel vocoders), FIR filters derived from homomorphic analysis (homomorphic vocoders), formant pole-pair filters derived by using formant analysis (formant vocoders), and many more. In all speech coders that use a vocal tract filter, the procedure is to analyze the original speech at the transmitter in order to extract the parameters that control the filter. These parameters are then encoded and transmitted to the receiver to control the vocal tract filter model used to generate the output speech signal.

A pitch-excited speech model represents the excitation signal as either a train

6 Introduction

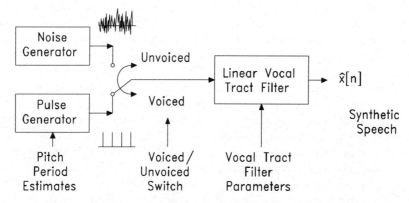

Figure 1.2. Pitch-excited vocoder synthesis model.

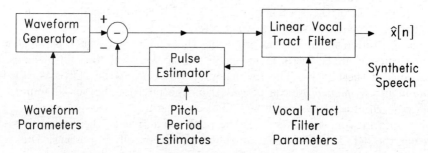

Figure 1.3. Waveform-excited vocoder synthesis model.

of pseudo-periodic pulses for voiced sounds or as a white noise signal for unvoiced sounds. The parameters of this excitation model are the pitch period for voiced sounds, the position of the voiced/unvoiced switch, and the excitation gain. Pitch-based excitation models are very compact and generally operate at bit-rates below 1000 bps for the excitation parameters. However, the pitch-excited model is inherently flawed in that it simply cannot represent some sounds (voiced fricatives) at all and, in general, cannot produce speech that is very natural sounding. In addition, the analysis procedures (pitch detectors) that must be used to determine the pitch excitation parameters are very difficult to realize effectively for a broad class of speakers and conditions. Thus, it is widely accepted that pitch-excited vocoding is not a good approach to achieve toll-quality speech where toll-quality speech means that the original and coded telephone speech signals are subjectively indistinguishable.

As an alternative to the pitch-excited model, the waveform-excited vocoder models the excitation signal as the output of a time-varying *pitch predictor* excited by a coded excitation signal. In this model, the pitch redundancy is modeled by the pitch predictor while aspects of the excitation signal that cannot be modeled by the pitch predictor are included in the coded waveform. Many different speech coders conform to this general model, including *adaptive-predictive coders*, *residual-excited vocoders*, *voice-excited vocoders*, *code-excited vocoders*, *multipulse-excited vocoders*, *self-excited vocoders*, and many others. In general, this entire class of vocoders requires more bits to represent the excitation signal, but is less constrained by the parametric model than the pitch-excitation model. In addition, because this class of vocoders does not require an explicit voiced/unvoiced decision and is less sensitive to pitch errors, the required pitch parameter estimation procedures are simpler to realize and are more effective than for the pitch-excited case. Waveform-excited vocoders are good candidate systems for low bit-rate, high-quality speech coders.

Aural Models A common feature of all speech coding systems is that they generate signals that are to be perceived by human ears. If there is any information in the original speech signal that is filtered out by the ear, then that information can be left out of the coded representation, and the required bit-rate can be reduced accordingly.

A model for the ear that has been used successfully by speech coding systems is the well-known *critical-band* model for aural perception. In the critical-band model, the ear is represented by a continuous bank of linear filters in which the bandwidth of the individual filters varies with frequency. Table 1.1 lists the characteristics of one set of experimentally determined critical bands that span a large fraction of the audible spectrum. The bandwidths and center frequency spacings are nonuniform and roughly correspond to a 1/6-octave filter bank. A simpler, five-band filter bank is illustrated in Figure 1.4.

The critical-band model tries to capture a number of related aspects of aural perception. The first is *aural frequency resolution*. The bandwidth of a critical band at a particular frequency is a measure of how distant in frequency two tones need to be in order to be distinguishable from one another. Simply, the ability of the human ear to resolve tones is roughly proportional to the width of the critical bands in the region of the tones. The second aspect is *aural noise-masking*. Stated simply, a (speech) signal in a particular critical band will mask another (noise) signal in the same band. Thus, noise signals that are close enough to speech signals in the frequency domain are masked, and noise signals that are not close enough to speech signals in the frequency domain are not. This model, in effect, specifies how the noise spectrum should be shaped as a function of the speech spectrum in order to achieve minimal perceptual impact.

Aural noise-masking models are used in two ways in modern speech coders: explicitly, in such frequency-domain coders as subband coders (SBC) and adaptive transform coders (ATC); and implicitly as a perceptual weighting function in such full-band coders as adaptive-predictive coders (APC), residual-excited coders, multipulse coders, and code-excited coders. Frequency-domain coders explicitly limit the coding noise to the band in which it was generated, thus exploiting the aural noise-masking effect. Full-band coders shape the quantization noise so that, to the maximum extent possible, it will be masked by the speech spectra.

Table 1.1. Critical band center frequencies and bandwidths.

Filter Number	Center Freq. (Hz)	Bandwidth (Hz)	Filter Number	Center Freq. (Hz)	Bandwidth (Hz)
1	50	70	13	1020	127
2	120	70	14	1148	140
3	190	70	15	1288	153
4	260	70	16	1442	168
5	330	70	17	1610	183
6	400	70	18	1794	199
7	470	70	19	1993	217
8	540	70	20	2221	235
9	617	86	21	2446	255
10	703	95	22	2701	276
11	798	105	23	2978	298
12	904	116	24	3276	321
			25	3597	346

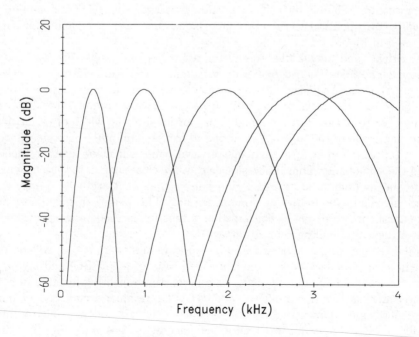

Figure 1.4. Critical band filters.

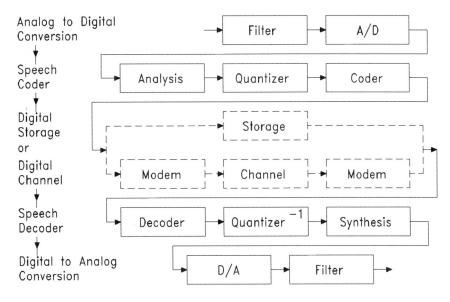

Figure 1.5. The elements of a speech coding system.

Comments on Models If viewed critically, the vocal tract models and the models of aural perception that have been used successfully in modern speech coders are definitely imperfect. In addition, the representation and parameterization of the models using DSP techniques are also not very good either. And yet, there is no question that they have led to major improvements in the performance of speech coding systems. The reason for this success is that the models have managed to capture, however poorly, some of the basic properties of speech production and speech perception that are important in speech coding systems. Future improvements in speech coding systems will undoubtedly come from a better understanding of speech production and perception, and from better DSP tools for exploiting such models.

1.2 Definition of a Speech Coder

All of the primary elements in a digital speech coding system are illustrated in Figure 1.5. At the left side of the figure is a list of the subsystems that are detailed within the boxes in the figure. The input to the system is a continuous speech waveform, $s(t)$. This signal is lowpass filtered using an anti-aliasing filter and sampled by an A/D converter, giving the digital speech signal, $s[n]$. This is the input to the speech coder.

The speech coder generally consists of three components: speech analysis, parameter quantization, and parameter coding. The input to the analysis stage is the digital speech signal, while the output is the new representation of the speech signal that will be quantized and coded. The output of the analysis stage may vary greatly depending on how the speech signal is being modeled. For a PCM system, for example, there would be no analysis at all because the output would simply be the digital speech signal. For other types of waveform coders, the output would be a processed version of the input. For parametric coders, the output would be the parameters of the speech model. Thus, for a pitch-excited LPC vocoder, the output of the analysis stage would be the linear predictive parameters of the vocal tract filter, the state of the voiced/unvoiced switch, the pitch period (if voiced), and the excitation gain. After analysis, these parameters are generally unquantized, and represent the best representation for the speech signal that can be attained by the analysis.

After analysis, the parameters must be quantized to reduce the number of bits required. In other words, quantization is used to reduce the intrinsic information rate of the speech representation. The output of the quantizer can be considered to be a noisy representation of the output of the analyzer. The output of the quantizer is provided to the coder which assigns a unique binary code to each possible quantized representation. These binary codes are packed together for efficient transmission or storage.

Digitally coded speech is often used in both communications applications and store-and-playback applications. These two classes of applications put very different requirements on the speech coding system. In a communications application, the system must minimize the coding delay, particularly when the channel may have other significant delays (such as in satellite communications). In addition, there is often a system cost and/or power constraint, particularly in high-volume consumer applications. This, in turn, constrains the computational intensity of the speech coding algorithm. Finally, real communications systems often introduce bit errors that must be addressed by the speech coder. To provide some protection against bit errors, a part of the available bit-rate is used, which leaves fewer bits to be used by the speech coder itself.

Voice storage applications typically have a different set of constraints. They usually do not have a constraint on the coding delay, and often the coder need not even be realized in real-time. In addition, they usually do not face a bit error environment. Thus, speech storage applications can often attain higher quality for the same bit-rate when compared to voice transmission applications. However, the system cost constraint is often more important than in a communications application, particularly for such applications as digital answering machines. In addition, the hardware cost for the speech coder in a voice storage application must be traded off against the cost of the digital storage itself.

The speech decoder reverses the operations of the speech coder. After the digital bit stream is decoded, it is transformed into quantized versions of the speech parameters through the inverse quantizer. In the absence of bit errors, this is identical to the parameters at the output of the quantizer stage in the speech coder. These parameters are then used to synthesize the coded speech signal, $\hat{s}[n]$. The synthesizer may be very simple, doing nothing at all for a PCM speech coder, for example. More often, it is quite complex because it realizes an entire parametric speech model for the system. Thus, for a pitch-excited vocoder, for example, it realizes the entire model of Figure 1.2. The resulting synthetic digital speech signal, $\hat{s}[n]$, is then D/A converted and passed through an anti-aliasing reconstruction filter to generate the analog synthetic speech signal, $\hat{s}(t)$.

Because this book is a laboratory text on speech coding, its emphasis is on the speech coder and speech decoder subsystems of Figure 1.5.

1.3 Classes of Speech Coders

As will be described throughout this book, modern speech coding systems typically take advantage of three classes of features in order to minimize the perceived coding distortion: the characteristics of the human auditory system, the characteristics of the vocal tract, and the characteristics of language and individual speakers. Typically, three separate characteristics of the auditory system can be used to advantage. These are (1) the aural noise-masking (critical-band) effect, (2) the frequency-variant sensitivity of auditory perception, and (3) the relative phase insensitivity of the ear (within critical bands). Likewise, a vocal tract model can be used in two different ways: either as the basis of a long-time stationary, time-invariant statistical model; or as the basis for a short-time stationary, time-varying model. If the short-time stationary model is used, then it may be applied in three ways: (1) by using syllabic energy variations, (2) by using an explicit slowly time-varying vocal tract model, or (3) by using the pseudo-periodicity (pitch) in voiced sounds.

The simplest common speech coding technique, linear pulse code modulation (PCM), makes essentially no assumptions (except as to dynamic range) about the characteristics of the signal being coded or the eventual use of the coded signal. As a result, linear PCM systems require the highest bit-rate to generate toll-quality speech. PCM systems have the advantage, however, that because they show no preference toward any particular class of signals, they may be used to code other classes of signals such as data signals or music. Companded pulse code modulation systems, such as the 64 kbps μ-law and A-law companded system used in telephone switching networks, make direct use of the ear's noise masking characteristics in their most basic form. In the resulting coded speech signals, the noise is correlated with the signal, which is good, but the noise also spreads throughout the frequency range with no regard to the presence of signal energy, which is bad. Companded PCM systems are extremely simple and inexpensive to implement, but the rate at which they must operate to achieve toll-quality speech is still relatively high.

Systems such as differential pulse code modulation (DPCM) and delta modulation (DM) make direct use of a long-term stationary statistical model for speech production. Adaptive differential pulse code modulation (ADPCM) and adaptive delta modulation (ADM) also make use of the slowly varying nature of the short-time energy, causing the noise to be heavily correlated with the speech signal, and causing a dramatic drop in the idle channel noise. Various forms of both the adaptive transform coder (ATC) and the adaptive predictive coder (APC) (some-

Coding Technique	Bit Rate (kbps)	Aural Noise Masking	Aural Freq.	Aural Phase	Syllabic Energy	Vocal Tract Model	Short Time Stationarity	Pitch
Linear PCM	80-120	No	No	No	No	No	No	No
Companded PCM	50-100	Yes	No	No	No	No	No	No
Delta Modulation	50-80	Yes	No	No	No	Yes	No	No
DPCM	40-80	Yes	No	No	No	Yes	No	No
ADM	16-40	Yes	No	No	Yes	Yes	No	Maybe
Subband Coder	10-32	Yes	Yes	Maybe	Yes	No	No	Maybe
ATC	8-32	Yes	Yes	Maybe	Yes	Yes	Yes	Yes
APC	8-32	Yes	No	No	Yes	Yes	Yes	Maybe
MPLPC	8-16	Yes	No	Yes	Yes	Yes	Yes	Yes
CELP	4-16	Yes	No	Yes	Yes	Yes	Yes	Yes
SEV	4-16	Yes	No	Yes	Yes	Yes	Yes	Yes
LPC Vocoder	0.6-2.4	Yes	No	Yes	Yes	Yes	Yes	Yes

Table 1.2. Summary of modern speech coding techniques.

times called the residual-excited vocoder) make use of all of the auditory and vocal tract features, and such systems are capable of generating excellent quality speech at medium to low bit-rates. The same is true of the newer parametric coders, such as the multipulse-excited linear predictive coder (MPLPC), the code-excited linear predictive coder (CELP), and the self-excited vocoder (SEV). Such systems were originally considered to be very complex and overly sensitive to both background noise and transmission errors. However, most of these early problems have been overcome, and many effective implementations now exist for this popular class of speech coders. Pitch-excited vocoders in general, and LPC vocoders in particular, are capable of operating at low to very low bit-rates, but they generally can never achieve toll-quality and do not perform well on either noisy speech signals or signals that do not contain speech. A summary of the speech features exploited by many modern speech coders is shown in Table 1.2.

1.4 Speech Coding Standards

Speech coding standards have played, and continue to play, an important role in the development and use of speech coders. There are a few speech coding applications in which interoperability is not an issue. A example of such an application is a digital answering machine or digital voice mail system in which the same system is used to both encode and decode the speech. For such applications, the speech coder of choice can be the best and most cost-effective one available at the time the system is designed, without regard to interoperability. For the vast majority of applications, however, interoperability is a major issue. All telecommunications applications clearly belong to this class, as well as all *carry-away media* applications, such as compact disks (CD) for music or speech. For interoperability to be achieved, standards must be defined and implemented.

Standards can be, and have been, developed by a number of different organizations. Perhaps the simplest standard that should be mentioned, although it is not specifically a speech coding standard, is the format used for digital CDs. Such systems use 16-bit linear PCM coding. As was discussed in the previous section, this format uses no features that are specific to speech, and thus is appropriate for coding both speech and non-speech signals.

Many of the earliest speech coding standards were created by the Department of Defense (DoD). This was primarily because digital encryption techniques are more effective than analog encryption techniques. One of the earliest standards still in use is a form of adaptive delta modulation called *continuously variable slope delta modulation* (CVSD), which can be used either at 16 kbps or 32 kbps. Although not a very good speech coder by modern standards, CVSD is reasonably good at 32 kbps and usable for military purposes at 16 kbps.

Two other DoD speech coder standards that are quite important are U.S. Federal Standards 1015 and 1016. The first of these, often referred to as *LPC10e* in its current form, is a pitch-excited linear predictive coder that operates at 2.4 kbps. The second and more recent DoD standard is a form of code-excited linear predictive coder (CELP) that operates at 4.8 kbps. CELP has certainly been the most popular coder of the 1980s, and this is reflected in the standards.

Another important standards body is the International Telecommunications Union (ITU), which is the successor to the International Telephone and Telegraph Consultative Committee (CCITT). This committee defines standards for the international telephone network. There are a number of important ITU standards currently in use. The most widely used are the 64 kbps companded PCM coders found in digital switching applications. They include the μ-law companded PCM for North America and the A-law companded PCM for Europe (ITU G.711). Other important telephone standards for 3.4 kHz speech signals include an ADPCM standard (G.726) operating at 16, 24, 32, and 40 kbps, and a low-delay CELP (G.728) operating at 16 kbps. There are also audio standards, such as a two-band subband coder (G.722) operating on 7 kHz signals at bit-rates of 48, 56, and 64 kbps.

The most active area of standardization at the time of this writing involves standards for digital cellular telephony. In North America, the standards organization is the Telecommunications Industry Association (TIA), and it has adopted a *full-rate* standard based on *vector-sum-excited linear prediction* (VSELP), which is a form of CELP. The Japanese Digital Cellular (JDC) standards organization has adopted a similar full-rate standard. In Europe, the standards organization is the global system for mobile telecommunications (GSM) subcommittee for the European Telecommunications Standards Institute (ETSI), and it has adopted a full-rate standard based on *regular pulse excitation with long-term prediction* (RPE-LTP), a form of MPLPC. Efforts are now underway for determining *half-rate* standards.

The purpose of this book is not to present the characteristics of any specific speech coder. Thus, for the most part, no standard speech coders will be discussed in detail. Rather, the purpose is to present the underlying principles on which these coders are based.

1.5 How to Use This Laboratory Text

This textbook, with its associated software, constitutes an introductory laboratory course on speech coding. It is not possible for a text of this size and scope to exhaustively present the field of speech coding, and this text does not pretend to do so. Rather, this book introduces the fundamental concepts that underlie modern speech coders and allows the reader to develop insight experimentally into how these concepts operate in various combinations.

This text accomplishes its goals by providing a diverse set of speech coding programs included in an interactive experimental environment that operates on an AT-compatible computer. Because many of the speech coding programs provided are quite computationally intense, a relatively powerful computer (a personal computer with at least an 80386 processor, a floating point coprocessor, and a mouse) is highly recommended. Also, because speech coders must be heard to be appreciated, a D/A system is recommended (see Chapter 2).

Chapter 2 describes the operation of the experimental environment and Chapters 3 to 8 describe the speech coders themselves. Each speech coder is described in three different ways. First, it is described in the body of the text using a compact mathematical description of its operation. Second, also in the body of the text, it is described in terms of the parameters that are required to control its operation. Finally, the implementation of each coder in the software environment has a context-sensitive help function that facilitates the experimental process.

This text does not need to be read from cover to cover. Chapter 2 outlines the installation and operation of the software environment. The reader should review this chapter in detail before using the software environment. After Chapter 2 the speech coder chapters should be read while using the experimental environment for support. It is recommended that readers maintain the results of their exercises (such as SNR values, etc.) because these results will often be compared with results from coding techniques introduced in later chapters.

Two different types of exercises can be found in the speech coder chapters. The *tutorial* exercises are in the body of the text, and are intended to support the textual materials. In addition, more general exercises are included at the end of each chapter. It is important to note that the software provided with this book is a completely general experimental environment for speech coding, and its use is certainly not limited to the exercises. Thus, the exercises should be considered only as guides to the use of the programs. Chapter 9 contains a collection of more in-depth exercises that are suitable for class projects.

This book is not meant to be used alone, but rather in concert with other materials. It is most effective when used as the experimental portion of a course that presents speech coding concepts in more detail. The reader need not be a speech coding expert to use this book, and a great deal can be learned by using the programs in a purely experimental style. However, much more can be learned in the context of a broader DSP and speech coding course.

DSPLAB: The DSP Laboratory Software

2.1 Introduction

DSPLAB is an interactive graphical shell that runs on MS-DOS-compatible computers. The DSPLAB environment includes all the programs necessary to experiment with the quantization, analysis, and speech coding techniques described in the remainder of this book. No programming is required to use these tools because the graphical user interface allows you to run a variety of experiments with different input parameters.

In addition to the speech coding algorithms and signal utilities, two of the most important features of this environment are the signal plotting capability and the signal record and playback facility. One-dimensional, three-dimensional, and density plotters are provided to aid in the understanding and visualization of speech coding. The signal record and playback facility permits subjective comparisons of coding experiments with different parameters.

This chapter is organized as follows:

- Section 2.2 describes how to install DSPLAB.
- Section 2.3 describes how to use the graphical environment.
- Section 2.4 describes the File, Edit, and Display menus.

2.2 Installing the DSPLAB Environment

This section describes the installation and general use of DSPLAB. To install DSPLAB, put the DSPLAB Disk 1 in Drive A (Drive B may be substituted for Drive A). Type **A:** and press RETURN to change to Drive A. With the **A:** prompt

on the screen, start the installation program with the command:
INSTALL
Installation is straightforward and automated. Once you start the INSTALL program, you will be asked about the destination disk, destination directory, swap drive, name of your editor, type of graphics adapter, and type of sound board installed on your machine. You should know the answers to these questions before you start your installation.

2.2.1 Destination Disk

The destination disk is the hard disk to which you want to install DSPLAB. The INSTALL program detects how many hard disks your computer has and displays a list of them. The default is to install on Drive C. You can choose any hard disk as the destination disk. Because DSPLAB takes up about 5 megabytes of disk space, you must have at least this much free space available on the destination disk.

2.2.2 Destination Directory

The destination directory is the directory into which you want to install DSPLAB. The default is to install into a directory named DSPLAB. You can enter another directory name as the destination directory.

2.2.3 Swap Drive

Ashell$^{©}$ is the main shell program required for DSPLAB. One of the requirements for using Ashell is that you have a drive available to which Ashell can temporarily store (i.e., swap) information as needed. This drive can be either a physical drive (such as C:) or a RAM drive configured in your computer's memory. The amount of space needed varies, but at least 500 kbytes should be available.

If you have at least 500 kbytes of extended or expanded memory in your computer that you can configure as a RAM drive, then we strongly encourage you to use a RAM drive as your swap drive because of its fast speed. If you do not have enough extended or expanded memory for a RAM drive, you can use any hard disk as a swap drive as long as about 500 kbytes of disk space is available.

If you want to use extended or expanded memory in your computer as a swap drive, you must configure this memory as a RAM drive before you install DSPLAB. If you use MS-DOS 5.0, this example shows how to use commands in your CONFIG.SYS file to configure one megabyte of extended memory as a RAM drive:
> device=himem.sys
> device=ramdrive.sys 1024 /e

2.2.4 Name of Your Editor

DSPLAB lets you use an editor for editing certain files. DSPLAB must know the complete name, including the extension (EXE, COM, or BAT), of the program you use to start your editor. If you have the location of your editor program in the PATH statement in your AUTOEXEC.BAT file, DSPLAB requires only the name (including the extension) you use to start your editor. Otherwise the complete path must be specified.

2.2.5 Type of Graphics Adapter and Monitor

DSPLAB works with these five graphics adapter and monitor combinations:

- VGA adapter with color display
- VGA adapter with monochrome display
- EGA adapter with color display
- EGA adapter with monochrome display
- CGA adapter with monochrome display

Select the adapter and monitor combination that your computer has.

2.2.6 Sound Board Installation

DSPLAB has built-in support for five sound boards for listening to encoded speech signals and for recording new signals. DSPLAB also provides a configurable batch file that can be used to interface to any unsupported D/A system (see Appendix B). The supported sound boards include:

- The Elf digital signal processing card from Atlanta Signal Processors, Inc. (ASPI)
- The Peachtree digital signal processing card from Atlanta Signal Processors, Inc. (ASPI)
- The TMS32010 signal processing card from Atlanta Signal Processors, Inc. (playback only)
- The Sound Blaster Pro sound card from Creative Labs
- The Sound Blaster 16 sound card from Creative Labs

While you do not have to have one of these cards to run DSPLAB, they are required if you want to record and playback signals. If you do not have any of these cards, select None. All of these cards should be installed according to their respective instruction manuals before installing DSPLAB. For the ASPI Peachtree, Elf, and TMS32010 cards, the installation program will create the necessary ADP.PAR file in the BIN directory. The Sound Blaster Card does not require the DSPLAB installation program to create any files, but it is assumed that the complete Sound Blaster environment has been installed according to the Creative Labs documentation.

2.2.7 Modifying CONFIG.SYS and AUTOEXEC.BAT

Near the end of the installation procedure, the INSTALL program checks your computer for these statements in your CONFIG.SYS file:

FILES=20
BUFFERS=20

If INSTALL finds the FILES and BUFFERS statements and they are each set to at least 20, then INSTALL does not alter your CONFIG.SYS file. If INSTALL finds either or both of these statements to be less than 20, it increases the number to 20. If INSTALL does not find a CONFIG.SYS file, it creates one for you with FILES and BUFFERS statements both equal to 20. INSTALL then ensures that your computer will know where to look for certain DSPLAB files. This is done through a PATH statement in your computer's AUTOEXEC.BAT file. For example:

PATH=C:\DSPLAB\BIN

If INSTALL does not find an AUTOEXEC.BAT file, it creates one for you and includes a PATH statement similar to the example. If INSTALL finds an AUTOEXEC.BAT file that has no PATH statement, it adds a PATH statement similar to the example. If INSTALL finds an AUTOEXEC.BAT file that has a PATH statement, it adds the information to the end of the *first* PATH statement it finds so your computer knows the location of the DSPLAB\BIN directory.

2.2.8 Verifying Installation

INSTALL creates the DSPLAB directory, under which you will find these subdirectories:

- BIN (contains the PC executable programs and the dsplab.bat file)
- HLP (contains the help files)
- MAC (contains the macros used to interface to the graphical environment)
- MNU (contains the menu files)
- SPC (contains the parameter collection files)

The ASPI_ENV.PAR file is an ASCII file, in the BIN subdirectory, that contains several parameters (environment variables) important to running DSPLAB. The ASPI_ENV.PAR file is used to store the answers to the installation questions and sets up other variables to point to the appropriate subdirectories. This file is required for DSPLAB to function properly. An example of this file is shown in Figure 2.1 for a typical installation to the C: drive.

2.3 Using DSPLAB

To start DSPLAB, first make sure you are in the DSPLAB directory. Start DSPLAB by entering:

DSPLAB

```
macro=c:\dsplab\mac
spcpath=c:\dsplab\spc
msupport=c:\dsplab\mnu;c:\dsplab\hlp
elfpath=c:\dsplab\bin
pchpath=c:\dsplab\bin
aprompt=DSPLAB $p$g
aswap=d:
BEEP=/B
editor=ed.exe
mblnk=2
overwrit=YES
signals=c:\dsplab
plotmet=x_time
sboard=SoundBlaster_16
video=VGA
```

Figure 2.1. A typical ASPI_ENV.PAR file that is created during the installation of DSPLAB.

This loads and runs the DSPLAB environment. After an initial message is displayed, DSPLAB displays the main screen as shown in Figure 2.2.

2.3.1 DSPLAB Menus and Screens

DSPLAB uses pull-down menus as shown in Figure 2.3 to present program selections. Many menu selections initiate a DSPLAB process. Other selections present you with parameter collection windows (see Figure 2.4) in which you enter data. Each parameter collection window has four buttons: OK, CANCEL, RESET, and HELP.

Click on OK (or press END) to accept the values and continue.

Click on CANCEL (or press ESC) to close the window and return to the current menu if you decide not to work within a parameter collection window.

Click on RESET (or F5) to reset the displayed values to their default settings. The default settings are also known as the standard parameters.

Click on HELP (or press F1) for context-sensitive help information. This help information should be used if you have any questions concerning input parameters and their valid numerical ranges.

While working with the DSPLAB environment, you may be presented with additional windows from which you can select data or entries. To select from an additional window, click the mouse on your choice. You can also use the RIGHT ARROW key to move your cursor to the additional window, where you then use the UP ARROW and DOWN ARROW keys to move to your choice. Press RETURN to complete your selection.

22 DSPLAB: The DSP Laboratory Software

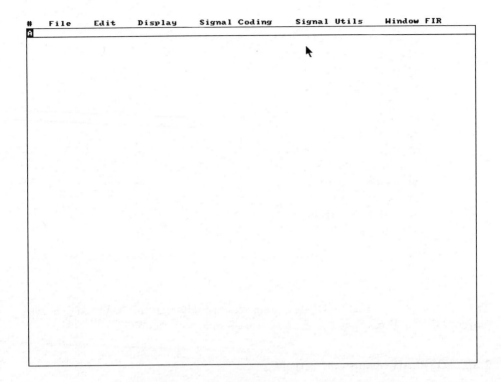

Figure 2.2. The initial DSPLAB main screen.

The *PopMenu* selection is available in many DSPLAB screens. Select *PopMenu* to return to the immediately previous menu.

The *File*, *Edit*, and *Display* menu selections are available from the DSPLAB main menu and every submenu and are described in Section 2.4.

2.3.2 Using the Keyboard

Although DSPLAB can work with a Microsoft-compatible mouse, you can also use the keyboard to select and enter information.

Menu Selections

To select pull-down menus, press *and hold* the ALT key, then press the first letter of the menu you want to select. For example, from the DSPLAB main menu, to select *Display*, press ALT and D. DSPLAB shows the *Display* pull-down menu. Use the UP ARROW and DOWN ARROW keys to move the cursor to the selection you want, then press RETURN. DSPLAB invokes that menu selection.

Press the ESC key to cancel a menu selection and return to the current menu. If

Sec. 2.3 Using DSPLAB 23

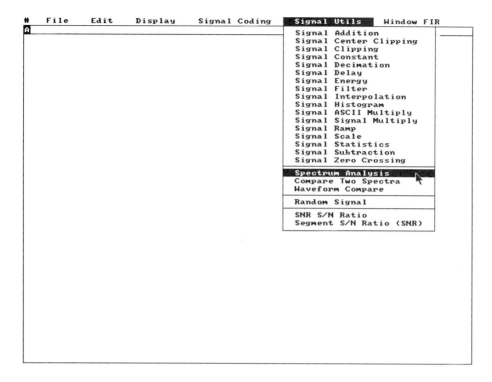

Figure 2.3. An example of a DSPLAB pull-down menu that shows the options under the Signal Utils menu.

you have a parameter collection window open, ESC closes the parameter collection window and returns you to the current menu.

Selecting and Entering Data

During operation, DSPLAB displays parameter collection windows (see Figure 2.4) in which you enter information in fields. Use the UP ARROW and DOWN ARROW keys to move to the fields in which you want to enter information. If information is already displayed, any data you enter writes over the old data.

For some fields, DSPLAB provides you with additional windows from which you can select data or entries. To select from an additional window, use the RIGHT ARROW key to move your cursor to the additional window, then use the UP ARROW and DOWN ARROW keys to move to your choice. Press RETURN to complete your selection.

Once you finish entering information in a parameter collection window, press the END key to end the parameter collection phase and subsequently begin the process.

24   DSPLAB: The DSP Laboratory Software

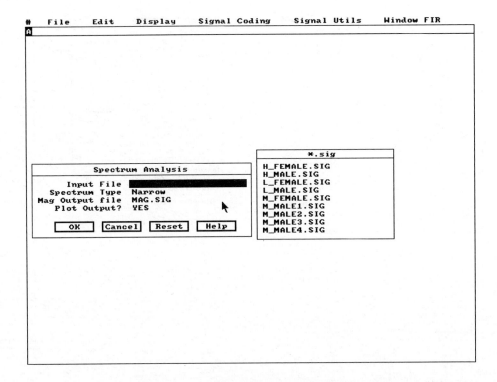

Figure 2.4. An example of a DSPLAB parameter collection window for spectrum analysis. The Input File parameter is blank, the Spectrum Type is set to Narrow, the Magnitude Output File is set to MAG.SIG, and the Plot Output switch is Yes, indicating the output will be plotted on the screen.

2.3.3 Using a Mouse

DSPLAB can work with a Microsoft-compatible mouse, letting you use the mouse cursor and buttons to select and enter information.

Menu Selections

To select pull-down menus, move the mouse cursor to the desired menu selection, then press *and hold* the mouse button. DSPLAB displays the pull-down menu. While holding the mouse button, drag the mouse cursor down to the selection you want, then release the button. DSPLAB invokes that menu selection. You can cancel a menu selection in two ways:

1. If, in the pull-down menu, you have not released the mouse button, move the mouse cursor off the pull-down menu, then release the button.

2. If you have released the mouse button and therefore displayed a parameter collection window, click on Cancel or press ESC. DSPLAB closes the window and returns you to the current menu.

Selecting and Entering Data

During operation, DSPLAB displays parameter collection windows (see Figure 2.4) in which you enter information in fields. Click the mouse on the fields in which you want to enter information. If information is already displayed, any data you enter writes over the old data.

For some fields, DSPLAB provides you with additional windows from which you can select data or entries. To select from an additional window, click the mouse on your choice in the additional window.

Once you finish entering information in a parameter collection window, click the mouse on OK to begin the process.

2.4 File, Edit, and Display Menus

This section provides a brief summary of the commands available from the File, Edit, and Display menus. More detailed information may be found at any time by clicking on the HELP button or hitting F1.

2.4.1 File Menu

The file menu is used for exiting the DSPLAB environment; changing the DSPLAB installation selections; performing file operations such as printing, deleting files, and recording and playing signal files; executing DOS commands; and saving display screens.

Configure DSPLAB

Configure DSPLAB lets you change the prompt used under Ashell, the swap drive, the state of the beep for error messages, the name of the editor, the number of menu blinks when selecting menu items, and the type of monitor connected to your PC.

Most importantly, this option lets you install a sound board to use as a D/A for playing out and recording speech signals. Six options are presently available for sound capability:

- ASPI Elf
- ASPI Peachtree
- ASPI TMS32010 (playback only)
- Creative Lab's Sound Blaster Pro
- Creative Lab's Sound Blaster 16

26   DSPLAB: The DSP Laboratory Software

- CUSTOM_D_to_A (playback only)

To use these boards, they must be installed into the computer according to the manufacturer's instructions and then installed into DSPLAB by selecting the proper Sound Board option.

Change Directory

Change Directory lets you change your current directory to another directory of your choice. The default is always the last directory in which you worked. When you select *Change Directory*, DSPLAB presents you with the *Change Directory* parameter collection window.

Delete File(s)

Delete File(s) lets you delete existing files. When you select *Delete File(s)*, DSPLAB presents you with the *Delete File(s)* parameter collection window.

Once you have selected the file or files to delete, press END or click the mouse on OK to delete the files. DSPLAB prompts you with:

Are you sure?

At this point, press Enter (or click on OK) to delete the files, or press ESC (or click on Cancel) to cancel the delete operation. If you choose to delete the files, DSPLAB displays:

File(s) have been deleted.

Memory Usage

Memory Usage displays a window that lists these categories and their current associated quantities of memory in bytes:

- Total amount of memory Ashell has allocated
- Heap size (the memory available to Ashell)
- Largest contiguous previously allocated memory block
- Amount of unallocated memory belonging to Ashell
- Amount of unallocated memory still belonging to DOS
- Total unallocated memory

Press Enter or click on OK to close the *Memory Usage* window. For proper operation, DSPLAB requires about 550 kbytes of free DOS memory. The amount of free DOS memory can be found by running either the DOS command MEM or CHKDSK. These DOS commands will display the amount of free DOS memory as either x *bytes free* or x *largest executable program size* where x should be larger than 550,000 for proper operation. If the amount of free DOS memory is not larger than this, consult your DOS manual for how to use your computer's extended memory to increase the amount of free DOS memory.

When using the *Memory Usage* command, if you find the value for the total unallocated memory is less than 50 kbytes, you should exit Ashell and then restart DSPLAB.

Save Screen

Save Screen saves the screen layout parameters to a file. This records the size of the windows, the number of windows on the screen, and the signals displayed. You can use this file and the *Restore Screen* menu selection to restore the screen.

Restore Screen

Restore Screen closes all display windows, erases all screen plots, and restores the screen from a file you created with the *Save Screen* menu selection. When you select *Restore Screen*, DSPLAB displays the *Restore Screen from File* parameter window.

If you created screen files with the *Save Screen* menu selection, you can select the screen file you want to restore. Use the mouse or arrow keys to make your selection, then press END or click the mouse on OK to invoke your selection.

Restore Screen Settings

Restore Screen Settings closes all display windows, erases all screen plots, and replaces the screen layout with the layout from a file you created with *Save Screen*. The windows open to the files you specify, not the files saved when you created the *Save Screen* file.

Print Signal Plot

Print Signal Plot sends the screen or window contents to your choice of a plotter or printer or to a disk file. When you select *Print Signal Plot*, DSPLAB displays the *Print Signal Plot* parameter collection window. DSPLAB supports these output devices:

- Canon BJ and LBP printers
- Epson 24-pin compatible printers
- Epson 9-pin compatible printers
- HP Laserjet compatible printers
- HP Paintjet compatible printers
- HPGL compatible plotters
- IBM Laser printers
- IBM Proprinter
- IBM Quietwriter 2 and 3
- NEC 24-pin compatible printers

- Postscript printers
- Texas Instruments 800 series printers

Support for all these output devices is provided by PrintGL. PrintGL is copyrighted shareware and must be registered if it is used beyond DSPLAB. Registration information may be found in the printgl.zip file provided in the bin directory of DSPLAB.

Playback Signals

Playback Signals lets you play signal files out the D/A of a supported sound board. Valid sound boards are the ASPI Elf, ASPI Peachtree, ASPI TMS32010, Creative Lab's Sound Blaster Pro, Creative Lab's Sound Blaster 16, and the CUSTOM_D_to_A driver. Once the sound boards have been installed according to their manufacturer's instructions, you may play up to 10 files sequentially. More information may be found in the help screen.

Record Signals

Record Signals lets you record new signals from the A/D of a supported sound board. Valid sound boards for recording are the ASPI Elf, ASPI Peachtree, Creative Lab's Sound Blaster Pro, and Creative Lab's Sound Blaster 16. More information may be found in the help screen. The record option is not supported with either the ASPI TMS32010 board or the CUSTOM_D_to_A driver.

Execute Ashell command

Execute Ashell command lets you temporarily exit DSPLAB to Ashell so you can issue Ashell commands or DOS commands. You can execute the Ashell command and up to eight arguments. Screen output (if any) from the executing command appears on the display.

DOS Shell

DOS Shell lets you temporarily exit DSPLAB to DOS (not Ashell) to issue DOS commands. *DOS Shell* adds a prefix (DOS.) to your computer's command line to show that you are operating with the DOS level. Type **exit** to return to DSPLAB.

Help Print

Help Print lets you extract a particular help screen to store as a disk file or to copy directly to a printer. The help file is stored as an ASCII file.

Quit DSPLAB

Quit DSPLAB discontinues operation of DSPLAB and returns you to the Ashell command line. To return all the way to DOS from Ashell, type **exit** and press RETURN.

2.4.2 Edit Menu

The *Edit Menu* allows you to edit ASCII files, inquire about and modify signal files, and playback and record signals.

Edit File(s)

Edit File(s) lets you edit an existing text file. When you select *Edit File(s)*, DSPLAB presents you with the *Edit File(s)* parameter collection window.

After you enter the information in the parameter collection window, press END or click the mouse on OK to have DSPLAB invoke the editor.

Signal File Information

Signal File Information displays information, such as file type and sampling rate, on a file that you choose. DSPLAB lists, in a window, all available signal files with a .SIG extension that can be located through the directory path specified by the SIGNALS environment variable. The SIGNALS environment variable is initialized in the ASPI_ENV.PAR file. *Signal File Information* displays this information for a signal file:

- File name
- File type
- Number of samples
- Revision
- Data type
- Data range
- Sample rate
- Descriptor
- Output device

The data range controls how the signal is normalized for display. Within DSPLAB the data range of the supplied signals is set to 1.0. This indicates that the signals are not normalized and therefore have amplitudes that range from $-32,768$ to $32,767$ (the full range of 16-bit integer samples). For all the applications in DSPLAB, the user does not need to be concerned with the data range.

Modify Signal File

Modify Signal File lets you alter any information, such as the sample rate and number of samples, displayed by the *Signal File Information* menu selection. DSPLAB lists, in a window, all available signal files with a .SIG extension that can be located through the directory path specified by the SIGNALS environment variable. You can then select the file to be modified.

You can convert a signal from the unnormalized signal plot to a normalized signal plot by changing its data range. If you force the data range of a 16-bit integer data file to be 16,384, then that integer signal will be plotted with amplitudes between -2.0 and 2.0; that is, the plotting routine will divide the signal values by the data range before plotting.

Convert Signal File

Convert Signal File lets you convert a signal file from one format to another. The five conversion modes are:

- ASPI to ASPI: changes formats and data range of signal files
- ASPI to BIN: changes ASPI SIG format to raw binary
- ASPI to ASCII: changes ASPI SIG format to ASCII
- BIN to ASPI: changes raw binary to ASPI SIG format
- ASCII to ASPI: changes ASCII to ASPI SIG format

If the input file is in ASPI format, *Convert Signal File* ignores the inputs for *Format* and *Range*. If the input file is in ASCII format, *Convert Signal File* ignores the *Input Format* field and assumes the *Data Type* to be floating-point. If the output file is in ASCII format, *Convert Signal File* ignores the *Output Format* field and assumes the *Data Type* to be floating-point.

Undo

Undo lets you undo the last signal cut, paste, or copy command that you have done.

Cut Signal

Cut Signal allows you to remove unwanted portions of signals, such as a noise pop or unwanted sounds. Upon selecting *Cut Signal*, a signal cursor will appear. Your next step is to position this cursor at the beginning of the portion of the signal to be cut. After clicking on the mouse, a second signal cursor will appear. This should be positioned at the last part of the signal to be cut. Clicking the mouse will remove this portion of the signal. *Undo* may be used to restore the signal portion if you have selected the wrong portion to remove.

Copy Signal

Copy Signal works the same as *Cut Signal*, only the signal portion is not removed from the input signal.

Paste Signal

Paste Signal allows you to duplicate portions of signals. The last selection that has been cut or copied is inserted after the position selected by the cursor.

Save Segment to File

Save Segment to File will save the signal that has been copied or cut to a signal file.

Play Segment

Play Segment is similar to *Playback Signals* except that it plays the currently selected segment of the waveform. More information may be found in the help screen.

Playback Signals

Playback Signals lets you play signal files out the D/A of a supported sound board. Valid sound boards are the ASPI Elf, ASPI Peachtree, ASPI TMS32010, Creative Lab's Sound Blaster Pro, Creative Lab's Sound Blaster 16, and the CUSTOM_D_to_A driver. Once the sound boards have been installed according to their manufacturer's instructions, you may play up to 10 files sequentially. More information may be found in the help screen.

Record Signals

Record Signals lets you record new signals from the A/D of a supported sound board. Valid sound boards for recording signals are the ASPI Elf, ASPI Peachtree, Creative Lab's Sound Blaster Pro, and Creative Lab's Sound Blaster 16. The record option is not supported with either the ASPI TMS32010 board or the CUSTOM_D_to_A driver. More information may be found in the help screen.

2.4.3 Display Menu

One of the most important features within DSPLAB is the plotting capability. You can create a display window layout, open and close display windows, and manipulate the amount and location of data displayed. There are three different types of signal plotters that can be used with DSPLAB. These are:

- Plot1d: a one-dimensional signal plotter
- Plot3d: a three-dimensional signal plotter

32   DSPLAB: The DSP Laboratory Software

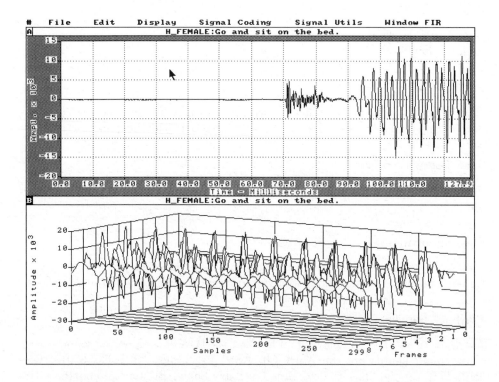

Figure 2.5. A speech signal plotted with the one-dimensional plotter and the three-dimensional plotter.

- Density: a density plotter

When opening a window to plot, you may select any of these types of plotters. The display data is always organized by frames. This allows for convenient display of framed analysis data. In addition, you can "lock" two or more displays together with the *Lock Windows* command, then scroll through all locked windows simultaneously. More detailed information is available in the help files. Figure 2.5 shows the output from both a one-dimensional and a three-dimensional signal plotter. In the rest of this section, the term "active window" is used to mean the window that is selected. You may select a window by clicking anywhere within it (see *Select Window* and *Tiled Screen*).

You can also save and restore the screen layout (a screen layout describes how the signal is plotted on the display). Each display window has six parameters:

- *File name*: the name of the signal file displayed in the window
- *Display length*: the amount of data displayed in the window
- *Current frame number*: used as the starting point for the data display

- *Frame length*: the length of a frame of data
- *Frame increment*: the number of frames to increment or decrement the currently displayed frame when you use the *Next Frame* or *Previous Frame* commands
- *Frame offset*: the number of samples the beginning of a frame is offset for display purposes

Locked Next Frame

Locked Next Frame shifts all locked windows to the next frame by the amount of the frame increment for each window. See also *Lock Windows*.

Locked Previous Frame

Locked Previous Frame shifts all locked windows to the previous frame by the amount of the frame increment for each window. See also *Lock Windows*.

Go to Frame n (Locked)

Go to Frame n (Locked) shifts the display in all locked windows to the frame number you specify.

Lock Windows

Lock Windows locks together the display windows you specify so you can manipulate the windows simultaneously with commands such as *Next Frame (Locked)*.

Next Frame

Next Frame shifts the displayed data to the next frame by the amount of the frame increment. For example, let's say you have on the screen a display with a display length of 512 samples, a frame size of 1 sample, a frame increment of 128, and the display goes from sample 0 to sample 511. Selecting *Next Frame* shifts the display "to the right" so it starts at sample 128 and ends at sample 639.

Previous Frame

Previous Frame shifts the displayed data to the previous frame by the amount of the frame increment. For example, let's say you have on the screen a display with a display length of 512 samples, a frame size of 1 sample, a frame increment of 128, and the display goes from sample 128 to sample 639. Selecting *Previous Frame* shifts the display "to the left" so it starts at sample 0 and ends at sample 511.

Open Multiple Windows

Open Multiple Windows allows you to clear the current screen and open from one to four signals with the same plot options.

Change Multiple Windows

Change Multiple Windows sets the display parameters for any of the display windows that you specify. For instance, you could change windows A and B by specifying AB as the window; or, by leaving this field blank, all the windows would change. This option should be used with the *Open Multiple Windows* command to change the way multiple windows are plotted.

Open Window

Open Window opens a display window and displays in it the signal file you select. This is similar to the *Open Multiple Windows* command.

Change Window Parameters

Change Window Parameters sets the display parameters for the display window you specify. The default window is the active window.

Close Window

Close Window clears the displayed information in the display window you specify and closes that window.

Close Current Window

Close Current Window clears the displayed information in the current window and closes the window.

Close All Windows

Close All Windows clears the displayed information in all display windows, then closes all display windows.

Redraw Screen

Redraw Screen redraws the screen and replots the signals.

Fast Display Menu

Fast Display Menu places the most frequently used *Display* menu items across the top of the screen (replacing the existing menu line) for faster and more convenient access. The menus present are:

- Display: same as *Display* except the following menu selections (normally on the *Display* menu but now displayed across the top of the screen) are not included:
- Next: same as *Next Frame*
- Previous: same as *Previous Frame*
- NxtLck: same as *Next Frame (Locked)*
- PrvLck: same as *Previous Frame (Locked)*
- Set: same as *Change Window Parameters*
- Goto: same as *Go To Frame n (Locked)*
- PopMenu

Selecting the *Fast Display Menu* replaces any other menus present. To return to the previous menu arrangement, select *PopMenu*.

Measure

Measure displays the signal values of the selected signal and sample. *Measure* activates a crosshair that you position, with the mouse or arrow keys, to the portion of the signal of interest. As you move the crosshair across a portion of the signal, DSPLAB displays a window of information that shows you values for that point's sample number, time, and level. Press ESC to cancel measure.

Zoom In

Zoom In lets you select and magnify a section of a displayed signal. When you select *Zoom In*, DSPLAB displays a plus sign (+). Use the mouse or arrow keys to move the plus sign to the beginning point of the section of the signal that you want to expand. DSPLAB also displays a window that shows you values for that point's sample number, time, and level. Information in this window lets you position the plus sign precisely.

Once you select the beginning point, click the mouse or press Enter to mark that point. Then move the plus sign to the end point. Click the mouse or press Enter to mark the end point. DSPLAB then magnifies the selected portion to fill the display window.

You can continue to zoom in on a portion of the displayed signal, or you can zoom out to the original display. To return to the display as it was before you magnified a portion of the signal, select *Zoom Out*.

Zoom Out

Zoom Out replots a signal that you magnified with *Zoom In* to its original displayed size.

Setup Windows

Setup Windows lets you select a screen layout for the display windows. You can plot one signal in each display window you create. You can select the display to be from full screen to four rows by four columns. Use the arrow keys or click the mouse on your selection of screen arrangements, then press END or click the mouse on OK to invoke your selection.

Select Window

For a screen with two or more display windows, *Select Window* lets you select a display window to be the active window. If you have only one window open, this command has no effect. You can also select any window to be the active window by clicking the mouse anywhere in that window.

Full Screen

For a screen with two or more display windows, *Full Screen* enlarges the active window to fill the screen. If you have only one window open, this command has no effect. You can also enlarge the active window to fill the screen by clicking the mouse on the letter (A, B, C, etc.) in the upper left corner of the window. Click the mouse on the letter again (or use the *Tiled Screen* command) to return the screen to displaying all windows in the tiled method.

Tiled Screen

Tiled Screen returns the screen to displaying all windows in the tiled method. If you have only one window open, this command has no effect. You can also reduce the active window and display all windows in the tiled method by clicking the mouse on the letter (A, B, C, etc.) in the upper left corner of the active full-screen window.

2.4.4 Freeze Display

One of the powerful features of the plotting system is the ability to freeze the screen layout while always plotting the most recent signal data on the screen. This means that the number of windows, their position, and their attributes are frozen, but the signal data is not. If you run any signal coding or utility function that modifies the signals that are displayed, the plots will be refreshed with the new signal data. This feature makes it easy to see the effects of experimenting with the parameters of the coders described in this text. To take advantage of this feature you should follow these steps:

1. The first time you run the coding experiment, select NARROW or YES for the *Plot Output?* parameter.
2. Use the *Locked Next* function or *Go to Frame n (Locked)* under the Display menu to move to an interesting frame of data. Frames of speech that contain

vowels (i.e., voiced speech) have significant signal energy (amplitudes on the order of 10,000) and are good candidates for seeing the effects of signal coding. Because some of the signal files included in DSPLAB start with silence (signal amplitudes on the order of 100), you will need to move farther into the signal file to accurately see the effects of the coding algorithm.

3. Rerun the coding experiment using the same output signal file name but modifying at least one of the coding parameters and select NO for the *Plot Output?* parameter. This will automatically update the plot of the output signal without resetting the display. This allows you to see precisely how the coding algorithm modifies your signal.

EXERCISE 2.4.1. **Using the One-dimensional Plotter**

Enter DSPLAB and clear all the windows by selecting *Close All Windows* from the Display menu. Select *Open Multiple Windows*, select Plot1D for the type of plotter, and select two of the provided signals to plot.

a. Enter a value of 128 for both the display length and the frame length. Enter 1 for the frame increment and 0 for the relative offset. Select x_time as the X-axis plot modifier and click on OK. The signals should be plotted on the screen with two windows, one above the other.

b. Select *Change Multiple Windows*, type AB for the windows to change, and change the display length and the frame length to 512. After clicking OK, you should see more of the signal data displayed in each of the windows.

c. Lock the two windows together with the *Lock Windows* command. Select AB for the window lock. Now select *Locked Next Frame*. Both signals will advance by one frame. Now compare this with selecting *Next Frame*. *Next Frame* will advance only the signal that had been selected (see *Select Window*).

d. Select *Change Multiple Windows*, type AB for the windows to change, and change the X-axis modifier to x_samples. After clicking OK, you should see the data plotted as discrete samples rather than as a continuous waveform. The X-axis is now labeled in samples rather than in milliseconds.

e. Select *Change Multiple Windows*, type AB for the windows to change, and change the display length and the frame length to 30,000. This allows you to see the entire signal. The portions with high signal amplitude are the voiced sounds and the other areas are either unvoiced or silent.

f. Click on the letter A in the upper left corner of the top signal window. This should enlarge the window to full-screen. Click on the letter A again. This should return the window to its original size and placement.

g. Under the File menu select *Restore Screen* and choose WIDE_2.SCR. This will restore a previously saved screen layout along with the signals that were being displayed at the time (see *Save Screen* in Section 2.4.1). DSPLAB has screen layouts prepared for comparing signals in either a Wide format (showing the entire signal) or a Narrow format (showing a single frame of the signal). The numeric suffix of the screen file indicates how many signals will be plotted.

h. Under the File menu select *Restore Screen Settings*. For the screen name select WIDE_2.SCR again and choose any two signal files. *Restore Screen Settings* will use the layout of the windows that was saved, but will use the signals that you selected instead of the saved files as in part **g**.

EXERCISE 2.4.2.   Playing a Signal File

Enter DSPLAB and under the Edit Menu select *Playback Signals*.

a. Using a sampling rate of 8 kHz, enter the filenames of up to 10 signals that you would like to playback. Make sure the analog output of your sound board is connected to a line-in connection on an audio amplifier. Click OK or press END. You should hear the sentences being played back.

b. Select *Playback Signals* again, only now enter a sampling rate of 16 kHz. When you playback the sentences they should sound high-pitched. By playing the signals back at higher sampling rates (compressing the signal in time), you expanded the frequency response of the signals. Try playing the signals at different sample rates.

EXERCISE 2.4.3.   Designing a Lowpass Filter

Enter DSPLAB and under the Window FIR Menu, select *Lowpass Filter*.

a. Using a sampling rate of 8 kHz, enter the specifications for a filter that approximates the response of a telephone line. The filter should have a passband cutoff at 3.2 kHz and a stopband cutoff at 3.6 kHz. Enter a ripple of 0.01, select a Kaiser window, and choose the calculated filter length. Click OK to design the filter. The signal plots that are displayed will include the Magnitude, Log magnitude, Impulse response, and Error signal.

b. Redesign the filter using a Hamming window with the same length computed for the Kaiser Window. Does the filter meet your specifications? If not, where does the response fail to meet your specifications?

c. Redesign the filter using a Kaiser Window and a passband cutoff of 1.8 kHz, a stopband cutoff of 2.2 kHz, and a ripple of 0.005. Choose a different name for the filter coefficients than the one used for part **a**. The result of designing this filter is a simple half-band filter; that is, a filter that retains half of the signal bandwidth and removes the other half. This filter is useful as either an interpolation filter or a decimation filter when changing the sample rate by a factor of 2.

d. Using the approach in part **c**, design a quarter-band filter. (A quarter-band filter will pass only a quarter of the available bandwidth.) Could this filter be used for interpolation or decimation? If so, what would be a good choice for the up or down sample factor associated with this filter?

EXERCISE 2.4.4. Filtering a Signal

Enter DSPLAB and under the Signal Utils menu, select *Signal Filter*.

a. Select an input signal (from the supplied signals) and an output signal. Use the filter that you designed in part **c** in Example 2.4.3 and plot the results. As described in Section 2.4.4, use the *Go to Frame n (Locked)* function to find an interesting portion of the signal. What is the main difference between the input and output signals?

b. Rerun part **a** with the same input and output signals but select the filter from Example 2.4.3, part **d**, and select NO to the Plot Output? question. The output signal that is displayed on the screen will be automatically updated with the new filtered signal.

c. Playback both the input and output signals from part **b**. Does the output signal sound different from the input signal?

EXERCISE 2.4.5. Interpolating a Signal

Enter DSPLAB and under the Signal Utils Menu, select *Signal Interpolation*.

a. Select an input signal from the supplied signals. Using the filter from Exercise 2.4.3, part **d**, enter the upsample factor associated with the quarter-band filter and the filter name. Does the output look like the input?

b. Select *Change Window Parameters* in the Display menu, and for window B enter new values for the display length and frame length that are the product of the upsample factor specified in part **a**, and the current display length and frame length, respectively. Does the output look like the input? Because of the interpolation, you have to change the number of points being displayed on the output signal in order to see the similarity.

c. Why is the output signal delayed with respect to the input signal? Try designing other lowpass filters with different lengths and see if the delay between the input and output signals changes.

EXERCISE 2.4.6. Spectral Comparison

Enter DSPLAB and under the Signal Utils menu, select *Spectrum Analysis*. This option will compute either narrow (fine frequency resolution) or wide band spectra of signals using an FFT length of 256 points. To enhance the display, a pre-emphasis filter is applied to the signal. This filter has a system response of $H(z) = 1 - 0.99z^{-1}$.

a. Select an input signal from the supplied signals, enter an output name, select Wide for the spectral type, and choose to plot the output. The response shows a one-dimensional plot of the input signal and the frequency response of the wide-band short-time Fourier transform.

b. Run the *Spectrum Analysis* option again with the same input signal. Select a different output name, Narrow as the spectrum type, and plot the display. Again, the output is a one-dimensional plot of both the input signal and the frequency response of the short-time Fourier transform. Notice that the spectrum is much more peaky than the first spectrum.

c. To compare the Narrow and Wide frequency plots, select *Compare Two Spectra* from the Signal Utils menu. Enter the two spectra names from parts **a** and **b** and select the density plotter. This will show the two frequency signals in side-by-side density plotters with frequency plotted along the abscissa and frames of data along the ordinate.

d. Compare the Narrow and Wide frequency plots using the plot3d plotter. Although the three-dimensional plotter is more useful for plotting LPC spectra, it does provide some useful visual information concerning the frequency content of the signal.

Quantization: PCM and APCM

3.1 Introduction

The simplest type of speech coder is one that directly samples the speech signal and then quantizes and codes the individual samples. Such systems make little or no use of the fact that they are coding speech signals, and their performance is generally not limited to speech signals. Such coders do have the capacity, however, to make some small use of vocal tract and aural models to achieve better performance. The effects they can use include the aural noise-masking properties of the ear and the syllabic energy variation of speech signals.

In general, a digital speech signal is obtained by sampling and quantizing a low-pass-filtered version of a continuous speech signal. Ideal sampling of a band-limited signal (i.e., sampling with infinite resolution at the Nyquist rate) is a process in which no information is lost. Digital speech is obtained by quantizing and coding the sampled speech signal, thus representing each sample with finite precision. Quantization is always a lossy process and distortion always occurs during quantization. Quantized speech can generally be modeled as a perfect, infinite resolution sampled signal plus a quantization error. The quantized signal is always distorted to some degree, and this distortion, if audible, is definitely not desirable in speech communication applications. The goal in the design of direct sample quantizers, therefore, is to design quantizers that result in the minimum possible perceived distortion in the coded speech at the target data rate.

In this chapter, we study the principles of the most commonly used quantization techniques. Quantizers are generally divided in two different ways: as *uniform* or *nonuniform* quantizers, and as *fixed* or *adaptive* quantizers. Uniform, fixed quantizers are the simplest type and will be considered first. Nonuniform and adaptive quantizers are more complex, but the added complexity is often worthwhile because

42 Quantization: PCM and APCM

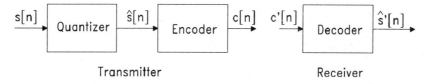

Figure 3.1. Block diagram of a pulse code modulator (PCM).

it can be used to reduce the perceptual effects of the quantization. In the systems considered here, the output of the A/D is assumed to be the original speech signal whose subjective quality is to be preserved. The original speech files available in the laboratory software are produced by a 16-bit A/D and thus have 2^{16} quantization levels. The quantization noise in these speech signals is essentially inaudible, but this representation requires a large number of bits to preserve the quality of the original speech waveform.

For historical reasons, a uniform quantizer followed by an encoder that assigns a code (usually a binary code) to each quantization level is called a *linear pulse code modulator*, or linear PCM. In such a linear PCM, the transmission bit-rate is

$$I = f_s B \quad \text{bits per second} \tag{3.1}$$

where B is the number of bits per sample and f_s is the sampling rate. For example, at a sampling rate of 8 kHz, the bit-rate is $8 \times 16 = 128$ kbps. Figure 3.1 shows a block diagram of a PCM system. The original input signal $s[n]$ is uniformly quantized and the quantized signal $\hat{s}(n)$ is encoded. The encoder assigns a binary code, $c[n]$, to each quantization level. The encoded signal is transmitted and the received code, $c'[n]$, is decoded in the receiver to obtain $\hat{s}'[n]$. Assuming a perfect transmission channel – that is, no bit errors due to the communication channel – $c'[n] = c[n]$ and the quantized signal $\hat{s}[n]$ can be reproduced exactly in the receiver – that is, $\hat{s}'[n] = \hat{s}[n]$. By replacing the uniform quantizer of Figure 3.1 with nonuniform or adaptive quantizers, *companding* PCM or *adaptive* PCM (APCM) systems are obtained. In companding PCM, speech is compressed before uniform quantization, whereas in APCM, adaptive quantizers are used.

Signal to Noise Ratio (SNR)

Nonuniform and adaptive quantization techniques use the properties of speech and the human auditory system in their simplest forms to improve the perceived quality of the coded speech. Both techniques attempt to preserve a high signal-to-noise ratio (SNR) for the coded speech. SNR is defined as the ratio of the signal energy to the quantization noise energy where, for zero-mean signals, the signal energy is defined as $\sigma_s^2 = \frac{1}{M} \sum_{n=1}^{M} (s[n])^2$ and the quantization noise energy is defined as $\sigma_q^2 = \frac{1}{M} \sum_{n=1}^{M} (\hat{s}[n] - s[n])^2$ where M is the number of points in the signal. Using a dB scale, SNR is defined by

$$\text{SNR} = 10 \log_{10} \frac{\sigma_s^2}{\sigma_q^2}. \quad (3.2)$$

SNR is an average measure of error over the whole speech signal and is usually dominated by the high-energy portions of the signal. An alternate measure of coding error is segmental SNR (SEGSNR), which is defined to be an average of SNR dB values obtained for smaller segments of the signal. Since the errors in low- and high-energy segments of speech are computed separately, perceptual quality is better reflected in SEGSNR than in SNR. The segment size is usually chosen to be on the order of the syllable duration (16–20 msec or 128–160 samples for 8 kHz sampled speech).

Nonuniform quantizers are typically used to produce a higher SNR for signals with a large dynamic range. Adaptive quantizers, on the other hand, attempt to produce digital signals with a high average SNR. In this chapter, uniform quantizers, nonuniform quantizers, and adaptive quantizers are discussed.

3.2 Uniform Quantization

Figure 3.2a shows the input-output function of a *mid-tread* uniform quantizer with eight quantization levels. In this figure, $s[n]$ is the input signal and $\hat{s}[n]$ is the output quantized signal. As shown in this figure, each quantization level is assigned a 3-bit binary code for encoding. In general, $B = \lceil \log_2 L \rceil$ bits are required to encode L quantization levels where $\lceil x \rceil$ represents the smallest integer larger than x. For example, encoding $L = 256$ levels requires 8 bits.

The process of uniform quantization is defined by representing all input values between s_{i-1} and s_i by $\hat{s}_i$ such that

$$s_i - s_{i-1} = \Delta \quad \text{and} \quad (3.3)$$
$$\hat{s}_i - \hat{s}_{i-1} = \Delta \quad (3.4)$$

where Δ is the quantization step size, and the quantization error, $q[n]$, is defined as

$$q[n] = s[n] - \hat{s}[n] \quad (3.5)$$

where

$$-\frac{\Delta}{2} \leq q[n] \leq \frac{\Delta}{2}. \quad (3.6)$$

A second type of uniform quantizer is known as the *mid-riser* quantizer and is shown in Figure 3.2b. The difference between mid-riser and mid-tread quantizers is that the mid-riser does not have a zero output level but is symmetric around zero. In contrast, the mid-tread quantizer has a zero output level but does not have the same number of positive and negative quantization levels. The parameters of a uniform quantizer, shown in Figure 3.3, are the number of quantization levels

44   Quantization: PCM and APCM

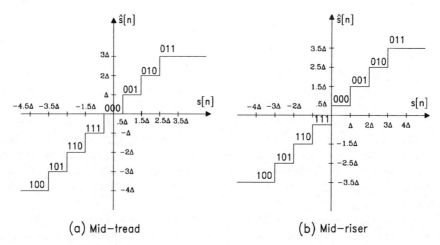

Figure 3.2. Mid-tread and mid-riser uniform quantizers with eight quantization levels. Each level is assigned a 3-bit binary code, $c[n]$.

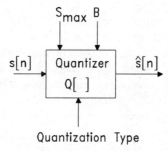

Figure 3.3. Block diagram of a uniform quantizer with its input parameters.

(or number of bits, B), the quantization step size, Δ, and the maximum signal amplitude, S_{max}. The quantization parameters are related by the equation

$$2S_{max} = \Delta 2^B \qquad (3.7)$$

where S_{max} is the maximum signal amplitude beyond which the signal is clipped; that is, set equal to the maximum signal value. If S_{max} is chosen to be small, then the signal will be clipped severely. On the other hand, a large S_{max} results in a large step size and, hence, a coarse quantization of the signal. The step size can also be reduced by increasing B. Increasing B results in a smaller quantization error but also a higher bit-rate for the coder. Table 3.1 shows the parameters of the uniform quantizer that are available in the uniform quantization program. This table also provides the range and some typical values for these parameters.

Sec. 3.2 Uniform Quantization

parameters	name	range	typical values
quantizer max. level	S_{max}	0–32767	32767
number of bits	B	1–16	7
quantization type	–	–	mid-tread or mid-riser

Table 3.1. The parameters of the uniform quantizer (PCM).

EXERCISE 3.2.1. **Bit-rate in PCM**

In this exercise, you will do some simple experiments that will demonstrate the behavior of uniform quantizers by changing the number of bits used to code each sample. Begin by finding the PCM Uniform coder on the Waveform menu (Signal Coding/Speech and Audio Coding/Waveform/PCM Uniform). Choose the mid-tread quantizer and set the maximum signal amplitude, S_{max}, to 32767.

a. In this experiment, you will run the uniform quantizer with 1, 3, 5, 8, 12, and 16 bits. For each case, you should save the output to a separate file, observe the resulting waveform graphically, and record the SNR and SEGSNR. The signals are best observed using the WIDE setting for the Plot Output? option. As the number of bits is increased, the coding error will decrease, the SNR will increase, and the coded and original waveforms will look very similar. (Note: The plot of the quantization error [window C] is automatically scaled, so you must observe both the error shape and the error scale.)

b. Make a plot of SNR versus the number of bits. What is the shape of this graph?

c. Listen to each quantized speech signal (File/Playback Signal) and find the minimum number of bits where the coded speech quality is as good as the original speech. This is referred to as *toll-quality* speech. What is the SNR in this case? What is the segmental SNR (SEGSNR)?

d. Using a narrow display window, compare your coded speech signals for 3, 5, and 12 bits. An easy way to do this is to use the narrow_3.scr screen template. First, create the basic display (File/Restore Screen Settings) using the narrow_3.scr template and your three coded speech files. Then, using the Display menu (Display/Next Frame Locked), step through 20 or 30 frames. What differences can you see in the coded waveforms?

e. At low bit-rates (1 to 3 bits/sample), compare the performance of the mid-tread quantizer to that of the mid-riser. Is there a significant difference between the two types of quantization? Is your answer still correct at higher bit-rates?

EXERCISE 3.2.2. **Step Size in PCM**

The purpose of this experiment is to investigate the effects of changing the step size with a fixed number of bits. Choose the mid-riser quantizer and set the number of bits to 6 (Signal Coding/Speech and Audio Coding/Waveform/PCM Uniform).

a. With the Plot Output? option set to WIDE, run the quantizer with the S_{max} values of 32767, 25000, 20000, 15000, 10000, and 5000. Save each output in a separate file. Observe the waveforms of both the original speech and the quantized speech for each case, and record the SNR.

b. Listen to each quantized speech signal. Recalling that $\Delta = S_{max}/2^{B-1}$, sketch a diagram of SNR versus step size, Δ. Find the step size value that results in the maximum SNR.

c. Find the maximum and minimum values of the signal that you used in parts a and b. You can use the Statistics function to compute the signal maximum and minimum (Signal Utils/Signal Statistics). You can make a good estimate of the best step size based on this information by setting S_{max} to the signal maximum. Since fine quantization requires a small step size, and hence a small S_{max}, there is a trade-off between the signal clipping distortion and quantization error. What is the SNR for your estimate?

d. To observe this trade-off more clearly, repeat the above experiments with a low number of bits, such as 2 or 3.

Since S_{max} is usually chosen to prevent severe clipping of loud signals, many of the higher quantization levels are not used for weaker sounds that have a lower magnitude. This results in a smaller number of effective quantization levels and hence a lower SNR for weaker sounds. To solve this problem, nonuniform quantizers with small step sizes at lower amplitudes and larger step sizes at higher amplitudes are used.

3.3 Nonuniform Quantization

Uniform quantizers perform optimally for signals with a uniform distribution over the dynamic range of the signal. As shown in the typical probability density function of a speech signal in Figure 3.4, speech signals do not have a uniform distribution.[1] As can be seen from this function, the speech signal is not uniformly distributed, and the majority of speech samples have relatively small magnitudes.

Another way to understand the basic problem associated with a uniform quantizer is to understand the time-varying properties of typical speech signals. Speech signals have syllabic energy contours, which means that the energy of the speech signal varies from large to small over syllabic intervals of a few tenths of a second.

---

[1]Computing the histogram of a signal can be done using the Histogram function from the Signal Utils menu.

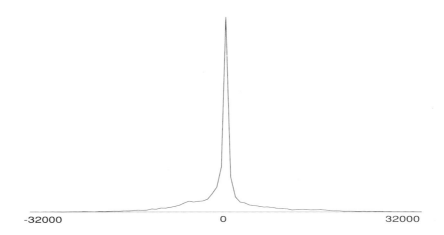

Figure 3.4. A typical density function for speech. This function is the histogram of 3 seconds of speech.

Figure 3.5 shows energy variation over a 500-msec speech segment. A fine uniform quantizer will produce uniform white noise at a constant level, which means the signal-to-noise ratio of the coded speech will vary at a syllabic rate, where it is worse during quiet periods (small signal, fixed noise) than in loud periods (large signal, fixed noise). We also know from the aural noise-masking property that, in an approximate way, signals mask noise. Thus a particular noise level will be more audible in quiet periods of the speech than in loud periods. Uniform quantizers result in fixed additive noise when what is really needed is a quantizer that results in something more like multiplicative noise; that is, low noise during low signal amplitudes and high noise during high signal amplitudes. Nonuniform quantizers whose step sizes increase with signal magnitude are thus more effective and can result in a relatively constant SNR over both high- and low-energy periods of the speech.

To address this need, Figure 3.6 shows the block diagram of a general *companding* (*comp*ressing-ex*pand*ing) PCM. In a companding PCM, the signal is passed through a nonlinear memoryless function, F, called a *compressor*. The compressor function, F, is chosen so that the compressed signal has a relatively uniform distribution. The compressed signal with a uniform distribution is then uniformly quantized and encoded. In the receiver, after decoding, the signal is *expanded* by applying the inverse function, F^{-1}, to the decoded signal.

48 Quantization: PCM and APCM

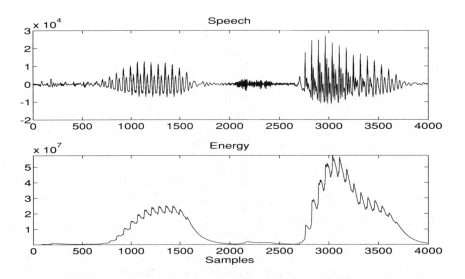

Figure 3.5. A 500 msec speech segment and its energy variation.

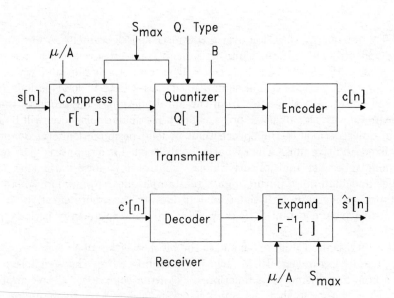

Figure 3.6. Block diagram of companding PCM. The function F is the compressor and F^{-1} is the expander.

Sec. 3.3   Nonuniform Quantization   49

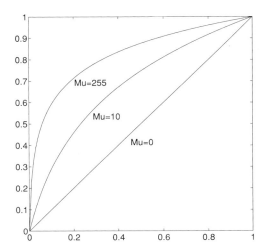

Figure 3.7. The μ-law compression function for three different values of μ.

3.3.1 Logarithmic Quantizers

The logarithm function is well suited to speech compression because it explicitly results in multiplicative quantization noise. Unfortunately, the logarithm function has a deficiency because it has an infinite range–that is, the logarithms of numbers approaching zero approach minus infinity. To avoid these problems, modified logarithmic functions are widely used in companding quantizers. Two common types of companders that are used in practice are μ-law and A-law companders. Both functions have logarithmic behavior and finite range. The μ-law compander is defined as

$$t[n] = F(s[n]) = S_{max} \frac{\log(1 + \mu \frac{|s[n]|}{S_{max}})}{\log(1+\mu)} sign(s[n]) \qquad (3.8)$$

where μ is the compression parameter, $sign(s[n])$ is the sign of $s[n]$, which has a value of ± 1, and log() is the natural logarithm function. Figure 3.7 shows the compression function for some typical values of μ. For small values of $s[n]$, this function is nearly linear because $\log(1 + \mu s[n]) \approx \mu s[n]$. Another logarithmic compressor is the A-law compander, defined by

$$t[n] = \begin{cases} \frac{As[n]}{1+\log A} & 0 \leq s[n] \leq \frac{S_{max}}{A} \\ S_{max} \frac{1+\log(A|s[n]|/S_{max})}{1+\log A} sign(s[n]) & \frac{S_{max}}{A} \leq s[n] \leq S_{max} \end{cases} \qquad (3.9)$$

where A is the compression factor. In A-law companding, the quantizer characteristics are linear for input magnitudes smaller than S_{max}/A and logarithmic beyond that.

50   Quantization: PCM and APCM

parameters	name	range	typical value
compression factor	μ	0–1000	255
	A	1-100	87.56
quantizer max. level	S_{max}	0–32767	32767
number of bits	B	1–16	5
quantization type	–	–	mid-tread or mid-riser

Table 3.2. The parameters of logarithmic companding PCM (log-PCM).

Notice that both μ-law and A-law companders map S_{max} to S_{max}. Hence, as in the uniform quantizer, S_{max} is the clipping level of both companding quantizers. Table 3.2 shows the parameters for μ-law and A-law companding nonuniform quantizers. The μ-law and A-law companders form the basis of the ITU $G.711$ speech standard for companding linear PCM samples. μ-law companders are used in North America with $\mu = 255$, and A-law companders are used in Europe with $A = 87.56$.

For minimum compression factors, $\mu = 0$ or $A = 1$, the function F is the unity function and no compression occurs. The larger values of μ or A result in compression and a constant SNR over a wider range of the input speech. Although very large values of the compression factors (e.g., $\mu > 500$ or $A > 100$) result in a constant SNR over a very wide range, they produce a lower overall SNR for the quantized speech.

EXERCISE 3.3.1.   Bit-rate in μ-law Companding PCM

In this exercise, you will do several experiments to observe the effects of companding on speech quantization. These experiments show how logarithmic and other companding quantizers can often outperform uniform quantizers.

a. Choose μ-law companding PCM and let $\mu = 100$ (Signal Coding/Speech and Audio Coding/Waveform/Log PCM Mu-Law). Set S_{max} to the real maximum value of the signal. The maximum value of the signal can be obtained using the statistics function under the Signal Utils menu. Run the μ-law quantizer with 1, 3, 5, 8, and 12 bits and save the results in separate files. In each case, record the SNR and SEGSNR and compare them to the SNR values obtained for the uniform PCM. Sketch the SNR versus bit-rate curve. Compare this sketch to that of the uniform quantizer. Do you achieve a better performance with the μ-law quantizer?

b. Listen to the signals, and compare the quality of each quantized speech signal to the corresponding signals generated using PCM with the same number of bits. What is the minimum number of bits for which the coded speech has *toll-quality*? What is the SNR in this case? What is the SEGSNR?

c. In this experiment, you will vary the value of μ and observe the results. From the Speech Coding menu, select Waveform and then choose μ-law companding PCM (Signal Coding/Speech and Audio Coding/Waveform/Log PCM Mu-Law). Using a 5-bit quantizer and a WIDE plot, create and save the coded speech signals for μ values of 0, 10, 100, and 255. Observe the compressed signal for different values of μ and record your observations along with the values for SNR and SEGSNR. Make a plot of SNR and SEGSNR versus μ. Using the Signal Histogram (Speech and Audio Coding/Waveform Coding/Signal Histogram) function of DSPLAB excluding zero, observe how the compressed signals have a more uniform distribution than the original speech signal. How would you characterize the effect of μ?

d. Choose a specific bit-rate. Use A-law companding instead of μ-law (Signal Coding/Speech and Audio Coding/Waveform/Log PCM A-Law). Compare the performance of the μ-law companding to that of the A-law companding in terms of SNR and subjective quality.

e. Based on the results from the above experiments, how would you compare SNR and SEGSNR for predicting the quality of companded quantizers?

3.3.2 Optimum Quantizers

Another approach to nonuniform quantization is to use *optimum* quantizers to maximize the average SNR. The goal of an optimum quantizer is to choose the nonlinear compression function so that for the given number of quantization levels, the resulting coded speech has the maximum possible SNR. The purpose of an optimum quantizer is to match the quantizer step sizes to the distribution of the signal being coded. Designing an optimum quantizer is equivalent to obtaining a function F_{opt} that results in a uniformly distributed compressed signal. The logarithmic quantizers approximate such a function using μ-law and A-law companders but are not optimal if the signal distribution is known. In cases where the signal distribution is known, it is possible to obtain the quantizer levels (or F_{opt}) that minimize the quantization error variance and maximize the SNR. To obtain the optimum quantizer, the quantization error variance given by

$$\sigma_e^2 = \sum_{i=\frac{-M}{2}+1}^{\frac{M}{2}} \int_{s_{i-1}}^{s_i} (\hat{s}_i - s)^2 p(s) ds, \qquad (3.10)$$

where $p(s)$ is the probability density function of speech, must be minimized. This can be done by differentiating σ_e^2 with respect to the quantizer parameters, s_i and $\hat{s}_i$ and setting the result to 0. The resulting optimum quantizer has the following properties:

1. The optimum location of the quantization level $\hat{s}_i$ is at the centroid of the probability density over the interval s_{i-1} and s_i.

52   Quantization: PCM and APCM

1 bit		2 bits		3 bits		4 bits		5 bits	
s_i	$\hat{s}_i$	s_i	$\hat{s}_i$	s_i	$\hat{s}_i$	s_i	$\hat{s}_i$	s_i	$\hat{s}_i$
∞	0.707	1.102	0.395	0.504	0.222	0.266	0.126	0.147	0.072
		∞	1.810	1.818	0.785	0.566	0.407	0.302	0.222
				2.285	1.576	0.910	0.726	0.467	0.382
				∞	2.994	1.317	1.095	0.642	0.551
						1.821	1.540	0.829	0.732
						2.499	2.103	1.031	0.926
						3.605	2.895	1.250	1.136
						∞	4.316	1.490	1.365
								1.756	1.616
								2.055	1.896
								2.398	2.214
								2.804	2.583
								3.305	3.025
								3.978	3.586
								5.069	4.371
								∞	5.768

Table 3.3. Optimum quantizer parameters for signals with Laplace density with 1, 2, 3, 4, and 5-bit quantization.

2. The optimum quantization boundary points lie halfway between the $M/2$ quantizer levels.

Speech has a density function that is relatively close to the Laplacian density function [5]. Table 3.3 shows optimum quantizer parameters for the Laplacian density for 1, 2, 3, 4, and 5-bit quantizers that can be used to implement the optimum Laplacian quantizer [6]. As an example, consider the 2-bit case in Table 3.3. For signals with unit variance, amplitudes between 0 and 1.102 are quantized to 0.395, and amplitudes above 1.102 are quantized to 1.810. The negative of these results applies to signal amplitudes that are less than zero. The optimum compressor characteristics for a signal with the Laplacian density function is known as an m-law compressor and is given as

$$t_{Laplace}[n] = S_{max} \frac{1 - e^{-m|s[n]|/S_{max}}}{1 - e^{-m}} sign(s[n]) \qquad (3.11)$$

where the optimum value of m is $m_{opt} = \sqrt{2} S_{max}/3\sigma_s$.

EXERCISE 3.3.2.   Optimum Quantizer

It has been shown that the Laplacian density closely approximates the speech density function. In this experiment, logarithmic quantizers are compared to the optimum Laplacian quantizers. To perform this experiment, you will use the General

Companding PCM function under the Waveform menu to perform a compression according to a piecewise linear function. The function accepts up to 100 points as samples of the desired compression function. The first and last points are assumed to be $(0,0)$ and (S_{max}, S_{max}) and therefore do not need to be specified. If no points are specified, the compression function is simply a linear function between $(0,0)$ and (S_{max}, S_{max}) and this results in no compression. For more detailed information, use the Help function for Signal Coding/Speech and Audio Coding/Waveform/General Companding PCM.

a. Use General Companding PCM (Signal Coding/Speech and Audio Coding/Waveform/General Companding PCM) with a simple compression function obtained by choosing the single point (3000, 15000). An example companding file containing this one point, called *compand.par*, is included with DSPLAB. You should also use the Edit function to examine the file format. Draw the compressor function and compare the SNR and SEGSNR you obtained by using this simple compressor to that of PCM. Do you observe a major difference? How might you come up with better compressor functions by using speech histograms?

b. For the simple companding function from part a, code a speech signal for 1, 3, 5, 8, 12, and 16 bits, save the results in separate files, and record the SNR and SEGSNR values. Listen to the results (File/Playback), and make a plot of SNR and SEGSNR versus bit-rate. How do these results compare to a μ-law quantizer?

3.4 Adaptive Quantization

In the previous section, we saw that nonuniform quantizers model the speech signal as a stationary signal with a wide dynamic range. The dynamic range must be wide to accommodate both loud and quiet segments of the speech. Because the speech signal is a nonstationary signal whose characteristics change in time, a better way to model the speech signal is as a nonstationary signal with a smaller dynamic range, but one in which the energy changes (relatively) slowly in time. A good way to address this nonstationarity and the high dynamic range of the speech signal is to use adaptive quantizers where the quantizer characteristics are dynamically changed to match the changing properties of the speech signal.

The basic idea in adaptive quantization is to vary the step size of the quantizer to match the variance of the input signal. Alternatively, and equivalently, the speech signal can be adaptively scaled to match the properties of a specific quantizer. Adaptation strategies can be based on either the input speech signal or the output quantized speech signal.

Figure 3.8 shows a block diagram of a system that adapts the quantizer based on the input signal. The adaptation parameter, $\Delta(n)$, is estimated from the input signal, and is then used to control the quantizer. Because the receiver cannot estimate this value, $\Delta(n)$ is also transmitted to the receiver where it is used to

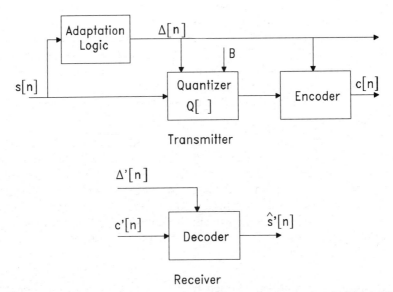

Figure 3.8. Block diagram of a feed-forward adaptive PCM (PCM-AQF).

control the decoding process. This system is called a *feed-forward* adaptive PCM (PCM-AQF) because the quantizer control parameters must be transmitted along with the encoded samples.

Figure 3.9 shows an alternative to the feed-forward approach where the system adapts based on the quantized output speech. Since the encoded speech is available at both the transmitter and the receiver, this system does not need to transmit the quantizer control parameters. For this reason, it is called a *feedback* adaptive PCM (PCM-AQB).

Depending on how often the adaptation is performed, adaptive systems can be divided into *instantaneous* and *syllabic*. In instantaneous adaptation, the quantizer parameters are updated at every sample of the input signal, whereas in syllabic adaptation, the changes are slower and are at a rate that is on the order of the syllable rate of speech (16–20 msec). In feed-forward systems, adaptation is usually performed at a slow (syllabic) rate because the adaptation parameters need to be transmitted after each parameter update. Conversely, since feedback adaptation does not require the transmission of any parameters, feedback adaptations are usually instantaneous.

3.4.1 Feed-forward Adaptation

In feed-forward adaptation, the input signal is used to estimate the adaptation parameters. A common approach is to use the speech variance in the adaptation process. The variance can be used in two different ways. First, it can be used

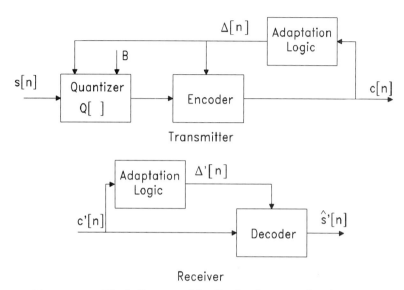

Figure 3.9. Block diagram of feedback adaptive PCM (PCM-AQB).

in a *step size adaptive* strategy where the step size of the quantizer is changed in time. In the second approach, the signal variance can be used in a *gain adaptive* strategy where the signal is scaled with a time-varying gain factor before being passed through a fixed quantizer. This process is often called *adaptive gain control* or AGC.

The variance of the speech signal in a feed-forward scheme can be estimated by computing the short-time energy of the speech signal, which can be defined as

$$\sigma^2[n] = \sum_{m=-\infty}^{\infty} s^2[m]w[n-m] \tag{3.12}$$

where $w[n]$ is a window function that is used to limit the amount of data or the duration of the speech used in the computation of the short-time energy. An exponential window, defined as

$$w[n] = \begin{cases} \alpha^{n-1} & n \geq 1 \\ 0, & otherwise \end{cases} \tag{3.13}$$

with $0 < \alpha < 1$ is commonly used. This results in the compact difference equation

$$\sigma^2[n] = \alpha\sigma^2[n-1] + (1-\alpha)s^2[n] \tag{3.14}$$

to compute the variance. Another common window is the rectangular window, which results in a variance estimate of the form

$$\sigma^2[n] = \frac{1}{M} \sum_{m=n}^{m=n+M-1} s^2[m] \tag{3.15}$$

56 Quantization: PCM and APCM

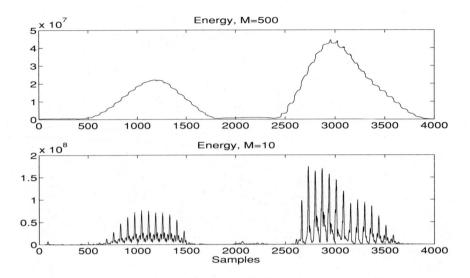

Figure 3.10. Variance estimates of a segment of speech with rectangular windows of length 500 (top) and 10 (bottom).

where M is the size of the rectangular window. While Equation 3.13 does not require any buffering of speech samples and therefore does not impose any processing delay, the computation of Equation 3.15 requires buffering M signal samples and consequently imposes M samples of coding delay. The estimate of Equation 3.15 is computed and applied for the speech samples between times n and $n + M - 1$.

The choice of α in the exponential window and M in the rectangular window controls the effective length of the interval that contributes to the variance estimate. While large values of α and M result in a syllabic variance estimate, small values result in an instantaneous estimate of variance. Figure 3.10 shows a syllabic variance and an instantaneous variance for the speech segment of Figure 3.5. The variances are obtained using rectangular windows with $M = 500$ and $M = 10$. Rectangular windows are usually used in feed-forward systems whereas exponential windows are preferred in feedback systems. With a syllabic adaptation, the side information is usually a small percentage (about 1%) of the total bit-rate [7].

Step Size Adaptation

Figure 3.11 shows the block diagram of a typical adaptive pulse code modulator (APCM) with feed-forward adaptive step size (PCM-ASF). In step size adaptation, the value of the step size is proportional to the variance of the speech signal. A good step size adaptation rule relates the step size to the estimated standard deviation

$$\Delta[n] = \frac{\Delta_0 \sigma[n]}{2^{n-1}} \tag{3.16}$$

parameters	name	range	typical value
number of bits	B	1–16	5
scaling step size	Δ_0	0.01–1	0.1
min. step size	Δ_{min}	0–500	50
max. step size	Δ_{max}	50–10000	500
window length (FF)	M	1–500	100
exponential window (FB)	α	0.1–0.99	0.95
quantization type	–	–	mid-tread or mid-riser

Table 3.4. The parameters of APCM with feed-forward (PCM-ASF) and feedback (PCM-ASB) adaptive step size.

where Δ_0 is a constant scale factor called the scaling step size. Using this adaptation strategy, the portions of the speech signal with high variance will have a large quantization step size, and the step size will decrease as the signal energy decreases. This results in an efficient use of available bits for different signal energy levels.

Table 3.4 shows the parameters of the PCM-ASF. The parameters include: the number of bits, B; the quantization type; minimum and maximum step sizes, Δ_{min} and Δ_{max}; the scaling quantizer step size, Δ_0; and the length of the rectangular window for the variance calculation.

Since the variance of the dynamic range of the speech signal may be very high, the adaptation in Equation 3.16 may yield unreasonably small or large step sizes. To prevent this, Δ_{min} and Δ_{max} are predetermined for the quantizer:

$$\Delta_{min} \leq \Delta[n] \leq \Delta_{max}. \quad (3.17)$$

The value of Δ_{min} is usually chosen to be small enough to minimize the idle channel noise, and Δ_{max} is chosen to be large enough to minimize the quantizer clipping. The ratio $\Delta_{max}/\Delta_{min}$ determines the dynamic range of the quantizer (with a typical value of 100), and Δ_0 determines the step size of the quantizer for a unit variance signal.

EXERCISE 3.4.1. PCM-ASF

In this experiment, the adaptation properties of an APCM with an adaptive feed-forward step size adaption (PCM-ASF) are evaluated, and its performance is compared to that of PCM and log-PCM. Two important issues in adaptive step size quantizers are the dynamic range of the step size and the speed of adaptation. In general, a very small Δ_{min} may result in a severe overloading of the quantizer in transition periods, and a large M will result in an unacceptably slow adaptation of the quantizer step size.

a. Run PCM-ASF (Signal Coding/Speech and Audio Coding/Waveform/APCM Step Size Adapt. Quant.) using the standard parameter values with a 3-bit

58 Quantization: PCM and APCM

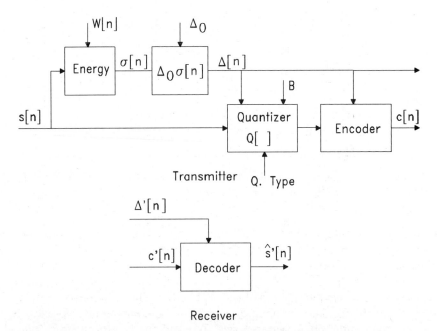

Figure 3.11. Block diagram of a feed-forward APCM with step size adaptation (PCM-ASF).

quantizer for values of Δ_0, the scaling step size, equal to 0.1, 0.5, 1.0, 2.5, 5.0, 10.0, and 20.0. Listen to the results and make a plot of SNR versus Δ_0. What value of Δ_0 gives the best quality? What value of Δ_0 gives the best SNR?

b. Run PCM-ASF (Signal Coding/Speech and Audio Coding/Waveform/APCM Step Size Adapt. Quant.) using the standard parameter values. Increase the dynamic range of the step size by decreasing Δ_{min} to small values and finally to zero in the sequence 25, 10, 2, and 0 with a fixed Δ_{max}. Observe the behavior of the step size for different values of Δ_{min}. How does the step size behave in the transition periods? How does the SNR change with Δ_{min}? How do the original and coded signals compare in high-energy regions? How do the original and coded signals compare in low-energy regions?

c. With a value of 1 for Δ_{min}, decrease Δ_{max} in the sequence 30000, 10000, 500, 100, and 50. Observe the behavior of the step size and the corresponding quantized speech as you change Δ_{max}. How does the SNR change with Δ_{max}?

d. Change the ratio $\Delta_{min}/\Delta_{max}$ using the set (100, 100), (25, 200), (10, 1000), and (1, 10000). Observe how the adaptation behaves and how the SNR changes with this ratio.

e. Using the results of previous experiments on PCM and log-PCM, compare the SNR (Signal Coding/Speech and Audio Coding/Waveform/Segment S/N Ratio

(SNR)) and the subjective quality of PCM, log-PCM, and PCM-ASF at similar bit-rates.

f. Starting from the standard parameter set and a 3-bit quantizer, investigate the effect of the window length M on the speech quality by trying values of M equal to 2, 50, 250, 1000, and 2500. What values of M give the best speech quality and SNR? Small values of M require a higher bit-rate. What is the smallest value of M for which smaller values do not give a quality improvement?

Gain Adaptation

Figure 3.12 shows the block diagram of a feed-forward gain adaptive APCM (PCM-AGF). In gain adaptive quantizers, the speech signal is scaled by a time-varying gain factor, $G[n]$, to obtain a quantizer input signal with a uniform range. The quantizer is fixed and its characteristics are chosen to match the scaled signal characteristics. The gain is chosen to be inversely proportional to the signal variance. Such a time-varying gain can be defined to be of the form

$$G[n] = \frac{G_0}{\sigma[n]} \tag{3.18}$$

where G_0 is a constant equal to the gain for the unit variance input. In this equation, a low signal energy results in a large gain and a high signal energy yields a small gain. This results in a signal with a relatively uniform range that is more suitable for a fixed quantizer than the original signal. To prevent overscaling of the signal, the variation of the gain function is usually limited by some maximum and minimum values such that

$$G_{min} \leq G[n] \leq G_{max}. \tag{3.19}$$

The ratio G_{max}/G_{min} controls the dynamic range of the scaled signal. By changing G_0, the signal can be uniformly scaled to any desired level. The parameters for PCM-AGF are given in Table 3.5.

EXERCISE 3.4.2. PCM-AGF

Feed-forward gain adaptation is very similar in many ways to feed-forward step size adaptation. Similar to PCM-ASF, the main issues in PCM-AGF are the dynamic range of the gain and the adaptation speed that is related to the window parameters. As in PCM-ASF, rectangular windows are used. The choice of G_0, G_{min}, G_{max}, and M results in different adaptation configurations.

a. Run PCM-AGF (Signal Coding/Speech and Audio Coding/Waveform/APCM Gain Adapt. Quant.) using the standard parameter values with a 3-bit quantizer for values of G_0 equal to 1000, 5000, 10000, 15000, and 20000. Listen to the results and make a plot of SNR versus G_0. What value of G_0 gives the best quality? What value of G_0 gives the best SNR?

60 Quantization: PCM and APCM

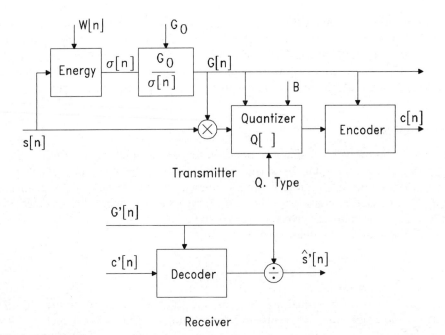

Figure 3.12. Block diagram of feed-forward gain adaptive PCM (PCM-AGF).

parameters	name	range	typical value
number of bits	B	1–16	5
quantizer max. level	S_{max}	0–32767	32767
scaling gain	G_0	–	–
min. gain	G_{min}	1–100	0.1
max. gain	G_{max}	10–1000	100
rectangular window length (FF)	M	1–500	100
exponential window parameter (FB)	α	0.1–0.9999	0.95
quantization type	–	–	mid-tread or mid-riser

Table 3.5. The parameters of feed-forward (PCM-AGF) and feedback gain adaptive PCM (PCM-AGB). A rectangular window is used in the feed-forward mode and an exponential window is used in the feedback mode.

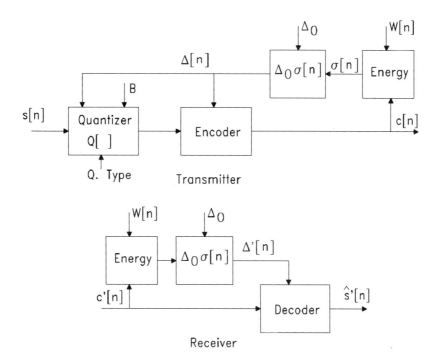

Figure 3.13. Block diagram of feedback step size adaptive PCM (PCM-ASB).

b. How would the experiments of part a change if the value of the signal maximum was reduced by half? Why?

c. Starting from the standard parameter set and a 3-bit quantizer, investigate the effect of window length M on the speech quality by trying values of M equal to 2, 50, 250, 1000, and 2500. What values of M give the best speech quality and SNR? Small values of M require a higher bit-rate. What is the smallest value of M for which smaller values do not give a quality improvement?

3.4.2 Feedback Adaptation

In feedback adaptive strategies, the adaptation parameters are estimated from the coded speech signal. Because this signal is available at the receiver, transmission of the side information, such as step size or gain, is not necessary. Figures 3.13 and 3.14 show the block diagrams of APCM with feedback adaptive step size quantizer (PCM-ASB) and feedback adaptive gain (PCM-AGB), respectively. Both systems use the adaptation rules of Equations 3.16 and 3.18.

In feedback systems, the estimate of the variance of the speech signal is based only on the past values of the coded signal. In the feedback mode, an exponential

62 Quantization: PCM and APCM

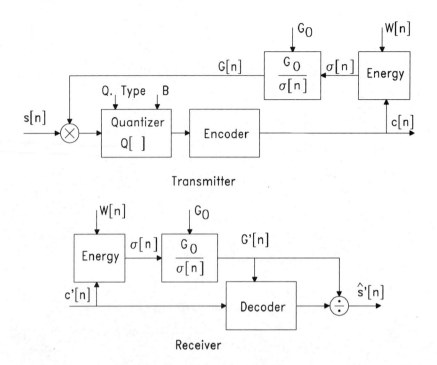

Figure 3.14. Block diagram of feedback gain adaptive PCM (PCM-AGB).

window is usually used that leads to the difference equation

$$\sigma^2[n] = \alpha \sigma^2[n-1] + (1-\alpha)\hat{s}^2[n] \tag{3.20}$$

to compute the variance. Notice that this estimate of the variance is based on the past values of the coded signal. The parameters of the feedback APCMs are given in Tables 3.4 and 3.5.

In feedback systems, to obtain a relatively good estimate of the signal variance, instantaneous adaptation is usually preferred. This is in contrast to the feed-forward case where slower adaptation is needed to reduce the amount of side information that must be transmitted to the decoder.

EXERCISE 3.4.3. **Feedback Quantizer**

In this exercise, the performance of a variance-based feedback adaptive quantizer is evaluated. In feedback quantizers, the variance is obtained from the coded signal and thus does not need to be coded and transmitted.

a. Use APCM with feedback adaption (Signal Coding/Speech and Audio Coding/Waveform/APCM Step Size Quant.) with the standard parameter set to code the talker of your choice with 1, 2, 3, 4, 5, and 6 bits. Make a plot of

B	PCM	DPCM
2	0.6, 2.2	0.8, 1.6
3	0.85, 1, 1, 1.5	0.9, 0.9, 1.25, 1.75
4	0.8, 0.8, 0.8, 0.8, 1.2, 1.6, 2.0, 2.4	0.9, 0.9, 0.9, 0.9, 1.2, 1.6, 2.0, 2.4
5	0.85, 0.85, 0.85, 0.85, 0.85, 0.85, 0.85, 0.85, 1.2, 1.4, 1.6, 1.8, 2.0, 2.2, 2.4, 2.6	0.95, 0.95, 0.95, 0.95, 0.95, 0.95, 0.95, 0.95, 1.2, 1.5, 1.8, 2.1, 2.4, 2.7, 3.0, 3.3

Table 3.6. Step size multipliers for 2, 3, 4, and 5-bit Jayant quantizers for PCM and DPCM.

SNR and SEGSNR versus bit-rate. Listen to your results. How many bits are required for *toll-quality*? It is generally believed that SEGSNR is a better estimator of subjective quality than SNR for systems with adaptive quantizers. Comment on this statement based on your results.

b. Using 3-bits and the talker of your choice, use APCM with feedback adaption with α, the feedback variance window parameter, and values of 0.999, 0.995, 0.99, 0.95, 0.9, and 0.5. Observe the step size parameter plot, make a plot of SNR and SEGSNR versus α, and listen to your results. What is the effect of α in terms of quality, SNR, and SEGSNR? Explain your results.

c. To compare the performance of the feed-forward and feedback adaptive quantizers, compare the results of PCM-AGF to those obtained in Exercise 3.4.2. Would you prefer a feedback or a feed-forward scheme at low bit-rates? How about at higher bit-rates?

Another common alternative for the feedback step size adaptation rule was studied by Jayant [6]. In this method, the step size of the uniform quantizer is adapted at each sample by scaling the step size by a variable step size multiplication factor $M[n]$, which is a function of the magnitude of the quantizer output at time $n-1$. The step size adaptation rule is

$$\Delta[n] = M[n]\Delta[n-1] \quad (3.21)$$

where $M[n]$ is the step size multiplier. Step size multipliers are usually provided in a table form. Some suitable values for step size multipliers are given in Table 3.6 for different numbers of bits. As in other adaptive step size schemes, limits are imposed on the step size so that

$$\Delta_{min} \leq \Delta[n] \leq \Delta_{max}. \quad (3.22)$$

As before, the ratio $\Delta_{max}/\Delta_{min}$ controls the dynamic range of the quantizer. Table 3.7 shows the parameters of the PCM-ASB with a Jayant quantizer.

64   Quantization: PCM and APCM

parameters	name	range	typical value
number of bits	B	1–16	4
min. step size	Δ_{min}	10–500	50
max. step size	Δ_{max}	100–10000	5000
step size multiplier	$M[n]$	0.1–10	0.7–3

Table 3.7. The parameters of a Jayant quantizer.

EXERCISE 3.4.4. **Jayant Quantizer**

The Jayant quantizer is a feedback APCM whose performance can be compared to other feedback APCMs. Some step size multiplier files are provided in DSPLAB. If all step size multipliers are set to 1, the quantizer becomes a PCM with a fixed step size.

a. Use APCM with the Jayant feedback adaption (Signal Coding/Speech and Audio Coding/Waveform/APCM Jayant Quantizer) with the standard parameter set to code the talker of your choice with 2, 3, 4, 5, and 6 bits. Since this is not a differential coder, use the step size multiplier files *mult2.ssm* to *mult8.ssm*. Make a plot of SNR and SEGSNR versus bit-rate. Listen to your results. How many bits are required for *toll-quality*? It is generally believed that SEGSNR is a better estimator of subjective quality than SNR for systems with adaptive quantizers. Comment on this statement based on your results.

b. Based on the results of the previous exercise, compare and contrast a Jayant feedback quantizer with an exponential feedback quantizer. How do they compare for SNR and SEGSNR? How do they compare in quality?

3.5   Exercises

EXERCISE 3.5.1. **Noise Analysis**

To understand the behavior of coding noise in the coders presented in this chapter, use the spectral analysis functions provided in DSPLAB under the Signal Utils menu. In a low bit-rate PCM, the noise is usually correlated to the speech signal, and is not white. However, at higher bit-rates, the coding noise is similar to white noise.

a. For 2 to 6 bits PCM (Signal Coding/Speech and Audio Coding/Waveform/PCM Uniform), analyze the coding noise. Listen to the coding noise and see if it is intelligible. You may use the Signal Scale function (Signal Utils/Signal Scale) to amplify the coding noise signal. Use the Statistics function (Signal Utils/Signal Statistics) to obtain some of the characteristics of the coding noise at different bit-rates.

b. Repeat part a for log-PCM (Signal Coding/Speech and Audio Coding/Waveform/Log PCM Mu-Law), and APCMs (Signal Coding/Speech and Audio Coding/Waveform/APCM Step Size Adapt. Quant.). Compare the performance of PCM, log-PCM, and APCMs based on your observations.

EXERCISE 3.5.2. Log-PCM

Logarithmic quantizers are known to be less sensitive to the ratio S_{max}/σ_s where S_{max} is a quantizer parameter and σ_s is the standard deviation of the signal. In this exercise, we measure this sensitivity for PCM (Signal Coding/Speech and Audio Coding/Waveform/PCM Uniform) and log-PCM (Signal Coding/Speech and Audio Coding/Waveform/Log PCM Mu-Law).

Choose a sentence and measure its standard deviation, σ_s. Run PCM with $B = 5$, and change S_{max} so that the ratio S_{max}/σ_s is between 1 and its maximum. Sketch a diagram of SNR (Signal Coding/Speech and Audio Coding/Waveform/SNR S/N Ratio) versus S_{max}/σ_s. Repeat this experiment for μ-law log-PCM. Compare the SNR diagrams. Which quantizer is more sensitive to the ratio?

EXERCISE 3.5.3. SNR at Different Ranges

In this exercise, the relatively uniform SNR of adaptive quantizers at different signal ranges is observed. The Signal Clipping function of DSPLAB allows clipping of the signal between two arbitrary values, C_1 and C_2.

a. Run PCM (Signal Coding/Speech and Audio Coding/Waveform/PCM Uniform) with standard parameters and 5 bits. Run the signal clipper (Signal Utils/Signal Clipping (2 sig.)) using the input and output signals from the PCM encoding as the two input signals for the clipper. Run the signal clipper with the values $(C_1 = 2000, C_2 = -2000)$, $(C_1 = 4000, C_2 = -4000)$, $(C_1 = 8000, C_2 = -8000)$, $(C_1 = 16000, C_2 = -16000)$, and $(C_1 = 32767, C_2 = -32767)$, where C_1 is the upper limit and C_2 is the lower limit. Sketch a diagram of SNR versus the range. Then repeat the entire experiment using center clipping (Signal Utils/Signal Center Clipping (2 sig.)) Explain your results.

b. Repeat part a for a Jayant (Signal Coding/Speech and Audio Coding/Waveform/APCM Jayant Quantizer) quantizer. How does the SNR (Signal Coding/Speech and Audio Coding/Waveform/SNR S/N Ratio) versus the signal range change? Use other adaptive quantizers and observe the results. Explain what you find.

Waveform Coding with Fixed Prediction

4.1 Introduction

The companding and adaptive quantization techniques discussed in the previous chapter begin to exploit the noise-masking property of aural perception and the syllabic energy variation of human speech signals. They do not, however, take advantage of the fact that the speech signal is produced by a human vocal tract.

To exploit the effects of the vocal tract, the speech signal can be considered from two perspectives: *long-term* and *short-term*. The long-term perspective considers the time-independent, average properties of speech and leads naturally to *nonadaptive* strategies for coding. The short-term perspective considers the slowly time-varying properties of speech caused by the mechanical properties of the vocal tract. This leads to a *short-time stationary* statistical model that is best exploited through *adaptive* speech coding strategies.

The theory of waveform representation based on nonadaptive *differential* coding is considered in this chapter. Differential coding refers to coding the difference between two signals rather than the signals themselves. The primary long-term property of speech that can be exploited by a nonadaptive differential coder is the relatively high correlation between adjacent samples. This is a result of the overall lowpass characteristics of speech due to the combined effects of the glottal pulse shape, the acoustic filtering effect of the vocal tract, and an acoustic radiation damping effect due to radiation from the mouth [3, 8]. Because of the high correlation between adjacent samples of the speech signal (especially for voiced speech), each speech sample can usually be estimated from the previous speech samples or coded speech samples. Further, because of the approximately linear properties of the acoustic vocal tract filter, the required estimation can be done with a linear filter [3, 9].

In differential coding, the short-term redundancy of the speech waveform is removed as much as possible, and the remaining signal is quantized and coded using techniques similar to those discussed in Chapter 3. This is accomplished by forming an *error signal* or *difference signal* by subtracting an estimate of the signal from the original signal. The estimate is generally obtained by a *linear predictor* that estimates the current speech samples from a linear combination of one or more past samples. Figure 4.1 shows the block diagram of this type of differential pulse code modulator (DPCM). The main source of performance improvement for differential coders is the reduced dynamic range of the quantizer input signal – that is, the difference signal $d[n]$ is generally smaller in amplitude than the original signal. Since the quantization noise is proportional to the step size, a signal with a smaller dynamic range can be coded more accurately with a given number of quantization levels.

The basic issues in differential coding are the predictor properties and the quantization of the prediction error signal. The simplest form of a differential coder is one that uses a fixed predictor and a fixed uniform quantizer. More sophisticated DPCMs include systems with adaptive quantizers (ADPCM),[1] adaptive predictors (APC), or both. Throughout this chapter, DPCMs with fixed predictors and different quantization techniques are presented. The prediction error signal can be quantized using any of the quantization techniques discussed in Chapter 3. Waveform coders based on adaptive predictors are discussed in Chapter 6.

4.2 Basic DPCM

Figure 4.1 shows the block diagram of a basic DPCM with a fixed predictor and a fixed uniform quantizer. In this figure, $\tilde{s}[n]$ denotes the estimate of $s[n]$ that is obtained from the previously coded samples as

$$\tilde{s}[n] = \sum_{i=1}^{P} a_i \hat{s}[n-i] \qquad (4.1)$$

where a_i, $i = 1, \ldots, P$ are the coefficients of the linear predictor $A(z)$, and $\hat{s}[n]$ are the previously coded speech samples. The predictor, $A(z)$, is defined as

$$A(z) = \sum_{i=1}^{P} a_i z^{-i} \qquad (4.2)$$

where P is the order of the predictor. The predictor coefficients are usually computed through a linear predictive analysis[2] of speech, where the a_i's are computed

---

[1] While the term ADPCM is often used in the literature to refer to coders that have adaptive predictors such as the ITU G.726 speech coding standard, within this book it denotes coders with adaptive quantizers and fixed predictors.

[2] Linear predictive analysis of speech is discussed in detail in the next chapter.

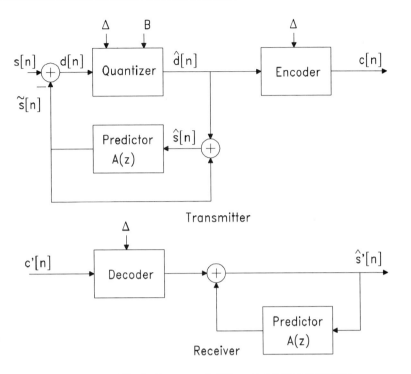

Figure 4.1. Block diagram of differential PCM (DPCM).

by minimizing the prediction error defined by

$$E = \sum_n (s[n] - \tilde{s}[n])^2. \tag{4.3}$$

The prediction error signal, $d[n]$, given by

$$d[n] = s[n] - \tilde{s}[n] \tag{4.4}$$

is the quantity that is quantized and coded. The quantized error signal can be represented as

$$\hat{d}[n] = Q(d[n]) = d[n] + q_d[n] \tag{4.5}$$

where $q_d[n]$ is the quantization error. As shown in Figure 4.1, the predictor input is

$$\hat{s}[n] = \tilde{s}[n] + \hat{d}[n]. \tag{4.6}$$

Combining Equations 4.4, 4.5, and 4.6, $\hat{s}[n]$ can be written as

$$\hat{s}[n] = s[n] + q_d[n] \tag{4.7}$$

which is the quantized signal. This shows that the output quantization error in this system is the same as the quantization error of the prediction error signal,

parameters	name	range	typical values
predictor order	P	1–10	1
max. quantizer level	S_{max}	0–32767	32767
number of bits	B	1–16	6
quantization type	–	–	mid-tread or mid-riser

Table 4.1. The parameters of the DPCM.

$d[n]$. This is significant because $d[n]$ has a smaller dynamic range (variance) than the speech signal, and consequently the corresponding quantization error will be smaller. The improvement in the SNR for DPCM compared to PCM is proportional to the *prediction gain*, defined by

$$G_p = \frac{\sigma_s^2}{\sigma_d^2} \tag{4.8}$$

where σ_s^2 is the variance of the speech signal and σ_d^2 denotes the variance of the prediction error signal, $d[n]$ [5]. It is obvious that a better prediction can improve the performance of DPCM because it reduces the variance of $d[n]$. Thus an important parameter of DPCM is its predictor, which is defined by the predictor order and predictor coefficients. Although increasing the order of the predictor usually results in a higher SNR, predictor orders higher than 4 or 5 yield marginal SNR improvements.

The quantization of the prediction error signal can be performed by any of the quantizer types discussed in Chapter 3. In the basic DPCM, a fixed uniform quantizer is used. The quantizer parameters are chosen to match the dynamic range and distribution of $d[n]$. Table 4.1 shows the parameters of the basic DPCM which includes the parameters of both the uniform quantizer and a fixed predictor. If the DPCM fixed predictor is properly designed and implemented, DPCM will result in measurably better performance in an SNR sense than PCM. As will be discussed in Chapter 6, even better performance is possible using adaptive prediction.

EXERCISE 4.2.1. DPCM

The advantage of differential coding in its simplest form is explored in this exercise. For this exercise, use the DPCM function (Signal Coding/Speech and Audio Coding/Waveform/DPCM Uniform Quantizer) of DSPLAB with a fixed uniform quantizer.

a. Using a 3-bit quantizer, implement a DPCM with a fixed first-order predictor, mid-tread quantizer, and Signal Max equal to 32767. A fixed first-order predictor with $a_1 = 0.85$ can be implemented using the *order_1.pcf* parameter file. Record the SNR and SEGSNR for your coded sentence, and listen to the coded speech. How good is it?

b. In DPCM, the difference between the input signal and the prediction signal, rather than the input signal, is coded. Hence, for best performance, the quantizer parameter (S_{max}) should be set to match the difference signal. The quantizer input difference signal (*qinput.sig*) is plotted as part of the DPCM plot function. Use the Statistics function to estimate S_{max}, and code the signal using this value. Compared to the speech signal, the prediction residual signal usually has a smaller dynamic range. Experiment with a few values around S_{max} (which, along with the number of bits, determines the step size) and find a step size that results in the highest SNR. This number is usually smaller than the step size used for PCM.

c. Repeat part a for 2, 4, 5, and 6 bits using the optimal S_{max} from part b. Sketch an SNR versus bit-rate diagram for DPCM. By comparing these results to those from uniform PCM, determine how PCM compares to DPCM based on SNR.

d. By listening to your results, find the approximate number of bits required to generate toll-quality DPCM coded speech.

EXERCISE 4.2.2. DPCM with First-Order Predictor

The values of the predictor coefficients in DPCM determine the effectiveness of the predictor in improving performance. In this exercise, you will measure the impact of the coefficient value in a first-order DPCM. The DPCM function reads the predictor coefficient file whose first row contains the predictor order, P, and the following rows contain the predictor coefficients: $a_1, a_2, \ldots, a_P$. These files have a *.pcf* extension, and you may create them using the Edit option of DSPLAB. *Order_1.pcf* is an example of such a file.

a. For this experiment, use a 4-bit uniform quantizer, and first-order predictor values of −0.99, 0.0, 0.5, 0.8, 0.9, and 0.99. To determine the SNR for each case, first run DPCM with the desired predictor value. Then, using the Statistics function, set S_{max} to match the maximum magnitude of the difference signal (*qinput.sig*) and run DPCM again. Listen to your results and make a plot of SNR versus predictor value. Remember that with a predictor value of 0, DPCM is equivalent to PCM. How sensitive is first-order DPCM to the choice of the predictor value?

b. The optimum fixed predictor for any of the sentences can be obtained in DSPLAB by using the Optimum Fixed Predictor function (Signal Coding/Speech and Audio Coding/Waveform/Optimum Fixed Predictor) which performs a linear predictive analysis on the whole sentence. Using this function, find the optimum first-order predictor for the sentence you used in part a. Using this predictor value, run the DPCM function again, and compare the optimum predictor value to the predictor values used in part a in terms of SNR.

c. Find the optimum first-order predictor value for five other talkers. How much variation do you find? What does this suggest about finding a single optimum predictor value for use for all talkers?

EXERCISE 4.2.3. **DPCM Predictor Order**

The optimum predictor for any of the sentences can be obtained in DSPLAB by using the Optimum Fixed Predictor function (Signal Coding/Speech and Audio Coding/Waveform/Optimum Fixed Predictor) which performs a linear predictive analysis on a whole sentence. In this exercise, you will study the performance of DPCM as a function of predictor order.

a. For a sentence of your choice, use the Optimum Fixed Predictor function to obtain optimum predictors of lengths $1, 2, 3$, and 10.

b. For each predictor order, determine the performance of DPCM in terms of SNR. To do this, first run DPCM with the desired predictor values. Then, using the Statistics function, set S_{max} to match the maximum magnitude of the difference signal ($qinput.sig$) and run DPCM again. Listen to your results and make a plot of SNR versus predictor order. How sensitive is the performance of DPCM to predictor order?

4.3 DPCM with Adaptive Quantization (ADPCM)

DPCM with fixed quantization provides an average of about 6 dB improvement in the SNR compared to PCM with the same number of bits/sample. As in APCM, adaptive quantizers can be used to obtain further improvement in coder performance. Figures 4.2 and 4.3 show the block diagrams of an adaptive DPCM (ADPCM) with a feed-forward adaptive quantizer (DPCM-AQF) and a feedback adaptive quantizer (DPCM-AQB). ADPCM may employ any of the adaptive algorithms described in Section 3.4. For example, the step size adaptation of Equation 3.16 can be used in both feed-forward and feedback systems. A possible approach in DPCM-AQF is to use the variance of $d[n]$ to adapt the quantizer step size. Since the variance of $d[n]$ is proportional to the variance of the original input signal, the adaptation can also be based on the input speech signal. In DPCM-AQB, the variance is estimated from $\hat{d}[n]$, the quantized samples of $d[n]$.

As given in Table 4.2, the parameters of ADPCM include the predictor parameters and the quantizer parameters. As in PCM-AQF, DPCM-AQF uses a rectangular window of length M for variance estimation. This variance is updated every M samples. In feedback adaptive ADPCM systems (DPCM-AQB), an exponential window with parameter α is used.

To obtain further improvement both in SNR and in subjective quality, the prediction gain of DPCM systems can be improved by using adaptive prediction techniques. DPCM with an adaptive predictor is discussed in Chapter 6.

EXERCISE 4.3.1. **DPCM with Adaptive Quantizer (ADPCM)**

As was seen in the previous chapter, adaptive quantizers are powerful tools for improving the performance of waveform coders. In this exercise, you will study the

Sec. 4.3 DPCM with Adaptive Quantization (ADPCM)

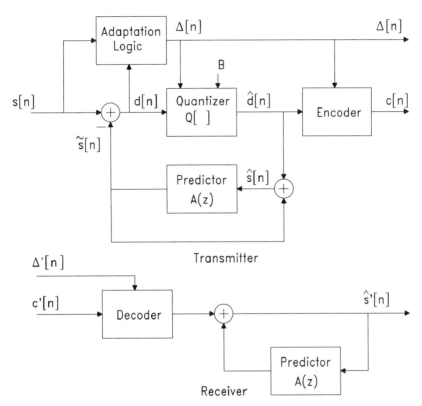

Figure 4.2. Block diagram of an ADPCM with feed-forward adaptive quantization (DPCM-AQF).

parameters	name	range	typical value
predictor order	P	1–10	5
rectangular window size	M (FF only)	1–500	50
exponential window	α (FB only)	0.1–0.99	0.95
number of bits	B	1–16	5
scaling step size	Δ_0	–	–
min. step size	Δ_{min}	50–500	100
max. step size	Δ_{max}	100–10000	5000

Table 4.2. The parameters of ADPCM. FF indicates feed-forward and FB indicates feedback.

74 Waveform Coding with Fixed Prediction

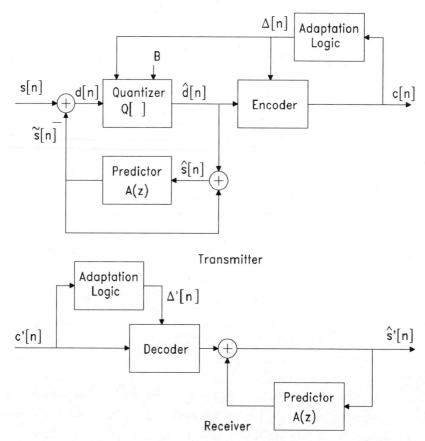

Figure 4.3. Block diagram of an ADPCM with feedback adaptive quantization (DPCM-AQB).

effects of combining DPCM with adaptive quantization by comparing ADPCM to APCM and DPCM.

a. Run ADPCM (Signal Coding/Speech and Audio Coding/Waveform/ADPCM Jayant Quantizer) with a Jayant quantizer at 2, 3, 4, and 5 bits/sample on the speech file of your choice. For this experiment, use a 1-tap predictor value of 0.85 (*order_1.pcf*). Different step size multipliers are needed for DPCM and PCM to obtain the best results. Use the step size multipliers for ADPCM that are given in Table 3.6 for this experiment. These step size multipliers can be found in control files *d_mult2.ssm* to *d_mult5.ssm*. Make a plot of SNR and SEGSNR versus bit-rate for ADPCM.

b. Listen to the results of part a. How does ADPCM compare to APCM and DPCM at the same bit-rates? At what rate is ADPCM toll-quality?

c. Make a plot of SNR and SEGSNR versus bit-rate that includes DPCM, APCM, and ADPCM. Which is more important for improving performance – adaptive quantization or differential coding?

d. For the ADPCM coding approach, which is a better measure of perceived quality – SNR or SEGSNR?

e. Repeat part a using the step size multipliers for APCM rather than ADPCM (step size multiplier control files *mult2.ssm* to *mult5.ssm*). How important are the step size multipliers to the quality of ADPCM?

4.4 Delta Modulation (DM)

Delta modulation (DM) is a specific type of DPCM with a first-order predictor ($P = 1$) and a 1-bit quantizer ($B = 1$). In DM, to compensate for the coarse quantization and to preserve the quality of the coded signal, the sampling rate of the DM input signal, f_{DM}, is chosen to be several times larger than the Nyquist frequency so that

$$f_{DM} = 2Rf_N \tag{4.9}$$

where R is the oversampling ratio and f_N is the Nyquist rate. Such a high sampling rate can be achieved by sampling the analog signal at the desired rate or, as is done here, interpolating the speech signal to obtain the desired sampling rate. The interpolation operation, which is shown in Figure 4.4, consists of an upsampling of the input signal followed by the interpolation filter. The output signal can be considered to be an oversampled version of the input signal. (See Chapter 2, Exercise 2.4.5, for an exercise on interpolation and decimation operations.) Such high sampling rates result in a very high sample-to-sample correlation in the sampled speech signal. This, in turn, results in a very small variance for the prediction error signal, $d[n]$, hence a high prediction gain, G_p. Figure 4.5 shows the block diagram of a DM. In DM, the predictor is of the form

$$p(z) = \alpha z^{-1} \tag{4.10}$$

where α is chosen to be smaller than 1. A predictor value of exactly 1 is usually avoided because it results in a full integrator synthesis filter, which tends to accumulate coding errors.

The main advantage of DM is its simplicity. In DM, the bit-rate is equal to the sampling rate of the DM input signal, f_{DM}. Obviously, the important parameters in the performance of DM are the oversampling ratio and the quantization type (either fixed or adaptive).

4.4.1 Linear Delta Modulation (LDM)

The linear delta modulator (LDM) is the simplest form of delta modulation where a fixed two-level quantizer is used. Figure 4.6 shows this 1-bit quantizer with step

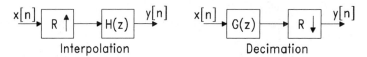

Figure 4.4. Block diagram of the interpolation and decimation operations. R is the up and down sampling factor, $H(z)$ is the interpolation filter, and $G(z)$ is the decimation antialiasing filter.

parameters	name	range	typical value
oversampling ratio	R	1–8	4
step size	Δ	100–1000	500
predictor parameter	α	0.1–0.99	0.85

Table 4.3. The parameters of linear DM (LDM).

size Δ whose output is defined by

$$\hat{d}[n] = \begin{cases} \Delta & \text{if } d[n] \geq 0 \\ -\Delta & \text{if } d[n] < 0. \end{cases} \quad (4.11)$$

LDM attempts to follow the input speech signal waveform in a linear fashion. The slope of LDM is given by the ratio Δ/T where T is the sampling period. Therefore, the slope can be controlled by changing Δ and T. Because reducing T results in a higher sample-to-sample correlation, the prediction gain can be increased by reducing T. The SNR can be increased by about 9 dB for each doubling of the oversampling ratio R, which corresponds to reducing T by half and doubling the bit-rate.

There are two types of distortion that are involved in DM: *granular noise* and *slope overload noise*. Granular noise is the result of oscillations of the DM output around the target value of the speech waveform. The step size value Δ needs to be small to reduce the granular noise. On the other hand, Δ needs to be large to prevent the slope overload that occurs when the slope of the signal is larger than the slope of the LDM [5]. Thus the proper choice of Δ is a compromise and is crucial for the performance of LDM. The parameters of LDM consist of the quantizer step size, Δ, the predictor coefficient, α, and the oversampling ratio, R. Table 4.3 gives some typical values for these parameters.

EXERCISE 4.4.1. Linear DM

The effects of granular noise and slope overload at different upsampling rates are studied in this experiment. In order to realize a DM system, there are three steps. First, the signal must be interpolated to a higher sampling rate. Second, the DM

Sec. 4.4 Delta Modulation (DM)

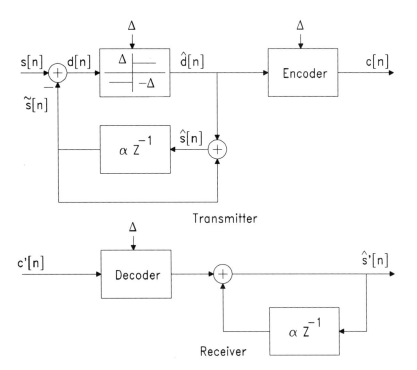

Figure 4.5. Block diagram of the delta modulator (DM).

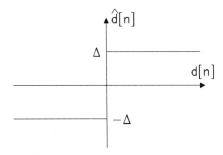

Figure 4.6. A 1-bit quantizer used in linear DM (LDM).

must be used to encode the interpolated signal. Finally, the signal must be downsampled to the original sampling rate for playback.

a. Using the Signal Interpolation function (Signal Utils/Signal Interpolation) and a sentence of your choice, create interpolated signals at a 1-to-2 and 1-to-4 rate. Recall the 1-to-2 interpolation requires a half-band filter and the 1-to-4 interpolation requires a quarter-band filter. The filter design function in DSPLAB (Window FIR/Lowpass) can be used to design the desired interpolation filter (see Chapter 2 and Exercise 2.4.3) or you can use the example filters (*1to2.flr* and *1to4.flr*) provided. Listen to your interpolated signals and explain what you hear.

b. Using your 1-to-2 interpolated signal as input, perform a linear DM using the LDM function (Signal Coding/Speech and Audio Coding/Waveform/LDM Delta Modulation) with a NARROW display for step sizes of 500, 2000, and 4000. Using the Display function, scroll through the signals and observe the original and coded signals in high-energy and low-energy regions. What happens in high-energy regions? What happens in low-energy regions? Explain.

c. Using your 1-to-4 interpolated signal as input, perform a linear DM using the LDM function with a NARROW display for step sizes of 500, 2000, and 4000. Using the Display function, page through the signals and observe the original and coded signals in high-energy and low-energy regions. What happens in high-energy regions? What happens in low-energy regions? Explain.

d. Using the Signal Decimation function (Signal Utils/Signal Decimation), downsample the delta modulated signals back to their original sampling rates and listen to the results. What is the effect of the quantizer step size on quality? How do 16 kbps and 32 kbps DM signals compare to PCM signals at the same rates?

4.4.2 Adaptive Delta Modulation (ADM)

Since the requirements on Δ to reduce both granular noise and slope overload are contradictory, an adaptive step size is desirable. In adaptive delta modulation (ADM), the step size is dynamically changed to satisfy both requirements. Such a time-varying step size can be used to prevent both granular noise and slope overload distortions by matching the slope of DM to the slope of the input waveform. The adaptive quantization principles discussed in Chapter 3 can be used in ADM. To preserve the simplicity of the coder and avoid transmitting any side information, feedback adaptation is usually preferred.

In this text, the system referred to as ADM has a Jayant-type quantizer whose step size adaptation rule is given by

$$\Delta[n] = M[n]\Delta[n-1] \qquad (4.12)$$

where $M[n]$ takes on one of the two values according to the rule given by

$$M[n] = \begin{cases} M_1 & \text{if} \quad c[n] = c[n-1] = c[n-2] \\ M_2 & \text{otherwise} \end{cases} \quad (4.13)$$

where $c[n]$ is the output code at time n. The condition $c[n] = c[n-1] = c[n-2]$ shows three consecutive increases or decreases of the signal. This requires a larger step size to prevent a severe slope overload. In general, M_1 and M_2 are chosen so that $M_1 > 1$, $M_2 < 1$, and

$$M_1 M_2 \leq 1 \quad (4.14)$$

for stability. As in other adaptive step size systems, minimum and maximum step size values, Δ_{min} and Δ_{max}, are used in ADM. Figure 4.7 shows a block diagram of an ADM system. As listed in Table 4.4, ADM is characterized by six parameters: the oversampling ratio, R; the value of the predictor parameter, α; the values of the quantizer control parameters, M_1 and M_2; and the minimum and maximum step size values, Δ_{min} and Δ_{max}.

parameters	name	range	typical value
oversampling ratio	R	1–8	4
predictor parameter	α	0.1–0.99	0.85
max. step size multiplier	M_1	1–10	1.2
min. step size multiplier	M_2	0–0.99	0.7
min. step size	Δ_{min}	10-500	20
max. step size	Δ_{max}	100-5000	2000

Table 4.4. The parameters of adaptive DM (ADM).

EXERCISE 4.4.2. **Adaptive DM**

Adaptive quantization can vastly improve the performance of a delta modulation system. The effects of granular noise and slope overload at different upsampling rates are studied in this experiment. Like a linear DM, there are three steps required to realize an ADM system. First, the signal must be interpolated to a higher sampling rate. Second, the ADM function must be used to encode the interpolated signal. Finally, the signal must be downsampled to the original sampling rate for playback.

a. Using the Signal Interpolation function (Signal Utils/Signal Interpolation) and a sentence of your choice, create interpolated signals at a 1-to-2 and 1-to-4 rate. Recall the 1-to-2 interpolation requires a half-band filter, and the 1-to-4 interpolation requires a quarter-band filter. The filter design function in DSPLAB (Window FIR/Lowpass) can be used to design the desired interpolation filter (see Chapter 2 and Exercise 2.4.3) or you can use the example filters (*1to2.flr* and *1to4.flr*) provided. Listen to your interpolated signals and explain what you hear.

b. Using your 1-to-2 interpolated signal as input, perform an ADM using the ADM function (Signal Coding/Speech and Audio Coding/Waveform/ADM Adaptive Delta Modulation) with a NARROW display for three different quantizer control pairs: (1.05, 0.8), (1.2, 0.6) and (2, 0.2). Using the Display function, scroll through the signals and observe the original and coded signals in high-energy and low-energy regions. What happens in high-energy regions? What happens in low-energy regions? Explain.

c. Using your 1-to-4 interpolated signal as input, perform an ADM using the ADM function with a NARROW display for three different quantizer control pairs: (1.05, 0.8), (1.2, 0.6) and (2, 0.2). Using the Display function, page through the signals and observe the original and coded signals in high-energy and low-energy regions. What happens in high-energy regions? What happens in low-energy regions? Explain.

d. Using the Signal Decimation function (Signal Utils/Signal Decimation), down-sample the delta modulated signals back to their original sampling rates and listen to the results. What is the effect of the quantizer control pairs on quality. How do 16 kbps and 32 kbps ADM signals compare to PCM signals at the same rates?

4.4.3 Continuously Variable Slope Delta Modulator (CVSD)

Another well-known DM is the continuously variable slope delta modulator (CVSD), which has a different adaptation rule for the quantization step size than ADM has. In CVSD, $\Delta[n]$ is computed as

$$\Delta[n] = \beta \Delta[n-1] + D[n] \qquad (4.15)$$

where $D[n]$ is equal to one of two constants, depending on whether the last three values of $c[n]$ were the same. The value of $D[n]$ is determined by

$$D[n] = \begin{cases} D_1 & \text{if } c[n-1] = c[n-2] = c[n-3] \\ D_2 & \text{otherwise} \end{cases} \qquad (4.16)$$

where $D_1 \gg D_2 > 0$. If the output increases or decreases for three consecutive samples, the step size is increased. Otherwise, the step size decreases at a rate specified by β. Values of β close to 1 result in slower variations of the step size. On the other hand, a smaller value of β and a large D_2 can be used for instantaneous adaptation.

Figure 4.8 shows the block diagram of a CVSD. In CVSD, the step size is inherently limited by the adaptation Equation 4.16. The *minimum step size* for CVSD is $D_2/(1-\beta)$, and the corresponding maximum step size is given by $D_1/(1-\beta)$. CVSD hence is characterized by five parameters: the input speech sampling rate; the value of the predictor parameter, α; the value of the *step integrator* parameter, β; the minimum step size parameter, D_2; and the corresponding maximum step size parameter, D_1. D_1/D_2 is known as the *expansion ratio* that determines the dynamic range of the quantizer. Table 4.5 contains the range of these parameters and some typical values for them.

parameters	name	range	typical value
oversampling ratio	R	1–8	4
predictor parameter	α	0.1–0.99	0.85
step integrator	β	0.1–0.99	0.95
step size parameter	D_1	1–1000	100
step size parameter	D_2	0–100	10

Table 4.5. The parameters of continuously variable slope DM (CVSD).

EXERCISE 4.4.3. CVSD

CVSD is just another form of adaptive DM. The primary difference between it and a Jayant ADM is the way in which the quantizer is controlled. The effects of granular noise and slope overload at different upsampling rates are studied in this experiment. Like a linear DM, there are three steps required to realize a CVSD system. First, the signal must be interpolated to a higher sampling rate. Second, the CVSD function must be used to encode the interpolated signal. Finally, the signal must be downsampled to the original sampling rate for playback.

a. Using the Signal Interpolation function (Signal Utils/Signal Interpolation) and a sentence of your choice, create interpolated signals at a 1-to-2 and 1-to-4 rate. Recall the 1-to-2 interpolation requires a half-band filter and the 1-to-4 interpolation requires a quarter-band filter. The filter design function in DSPLAB (Window FIR/Lowpass) can be used to design the desired interpolation filter (see Chapter 2 and Exercise 2.4.3) or you can use the example filters (*1to2.flr* and *1to4.flr*) provided. Listen to your interpolated signals and explain what you hear.

b. Using your 1-to-2 interpolated signal as input, perform an adaptive DM using the CVSD function (Signal Coding/Speech and Audio Coding/Waveform/CVSD Continuously Variable Slope DM) with a NARROW display for three different quantizer control parameters (β): 0.99, 0.80, and 0.50. Using the Display function, scroll through the signals and observe the original and coded signals in high-energy and low-energy regions. What happens in high-energy regions? What happens in low-energy regions? Explain.

c. Using your 1-to-4 interpolated signal as input, perform a CVSD using the CVSD function with a NARROW display for three different quantizer control parameters (β): 0.99, 0.80, and 0.50. Using the Display function, page through the signals and observe the original and coded signals in high-energy and low-energy regions. What happens in high-energy regions? What happens in low-energy regions? Explain.

d. Using the Signal Decimation function (Signal Utils/Signal Decimation), down-sample the delta modulated signals back to their original sampling rates and listen to the results. What is the effect of the quantizer control pairs on quality? How do 16 kbps and 32 kbps CVSD signals compare to PCM signals at the same rates?

4.5 Exercises

EXERCISE 4.5.1. DPCM with Optimum Predictor and Quantizer

To understand the limitations of the PCM and DPCM coders for coding music signals, run PCM, log-PCM, and DPCM algorithms on a music signal. Listen to the outputs and measure the SNR (Signal Coding/Speech and Audio Coding/Waveform/SNR S/N Ratio) and SEGSNR (Signal Coding/Speech and Audio Coding/Waveform/Segment S/N Ratio (SEGSNR)). Compare the results on music to your previous results on speech for each coder.

EXERCISE 4.5.2. Other ADPCM Systems

Two other ADPCM functions, one with feedback and one with feed-forward variance-based adaptive step size quantizers (Signal Coding/Speech and Audio Coding/Waveform/ADPCM Gain Adapt. Quant.), are provided in DSPLAB. Run these coders at 2, 3, 4, and 5 bits/sample and compare your results to the output of DPCM with a Jayant quantizer.

EXERCISE 4.5.3. ADPCM with Feed-Forward Adaption

In ADPCM with feed-forward adaptation, adaptation block size, M, is an important factor in performance and bit-rate. A small M results in a better adaptation while increasing the bit-rate. Starting from a relatively large M, find the value of M below which the SNR and quality do not improve significantly. Sketch an SNR versus M diagram.

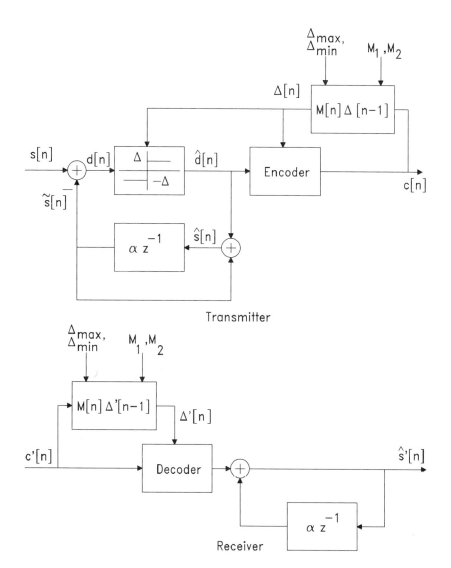

Figure 4.7. Block diagram of adaptive DM (ADM).

84 Waveform Coding with Fixed Prediction

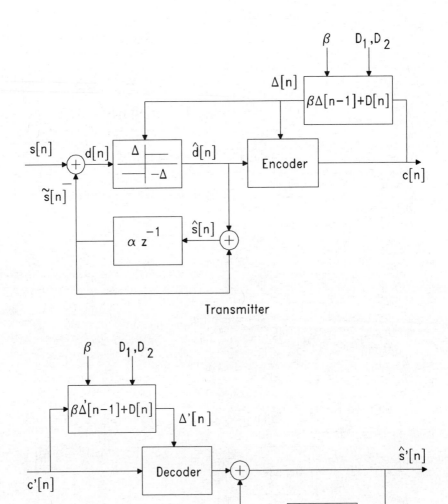

Figure 4.8. Block diagram of continuously variable slope DM (CVSD).

Pitch-excited Linear Predictive Vocoder

5.1 Introduction

Linear predictive coding (LPC) is one of the most popular coding techniques for speech signals and it has received extensive attention over the past two decades. LPC is not so much a type of speech coder as it is a technique that is used in a variety of different types of speech coders. It can be (and is) used in pitch-excited vocoders, voice-excited vocoders, waveform coders, analysis-by-synthesis coders, and even frequency-domain coders. This chapter will introduce LPC in the context of pitch-excited vocoders. Later chapters will address other classes of LPC speech coders.

The pitch-excited LPC has played a major role in the evolution of speech coders because it was the first type of LPC used for speech. Pitch-excited linear predictive coders also have the advantage that they are able to operate at low bit-rates using relatively modest computational resources while providing a very usable coded representation of the original speech signal. Their primary disadvantage is that the pitch-excited model limits the ultimate quality of the coder. Stated simply, pitch-excited LPCs can make good quality speech at low and even very low bit-rates, but they can never make toll-quality speech regardless of the number of bits employed.

It is not an exaggeration to say that linear predictive analysis is the single most important analysis-technique development in speech coding in the last 25 years. There are really three basic reasons for this. First, the linear predictive model is a good, albeit imperfect, discrete-time (digital) model for the acoustic vocal tract filter (see Figure 1.2). Second, at times when the linear predictive model is not a good match to the speech generation system, it still does a credible job of capturing the perceptually important speech properties. Finally, linear predictive analysis techniques are largely time-domain techniques, which makes them a good match to modern DSP microprocessors and other VLSI implementation techniques. Linear predictive analysis has had a similar impact in other speech processing areas. Other applications of linear predictive analysis include speech spectral analysis, vocal tract area function estimation, formant frequency and bandwidth estimation, and pitch prediction.

This chapter will introduce LPC techniques in the context of the pitch-excited LPC. These same basic LPC techniques will then be employed throughout the rest of the book in other classes of speech coders.

5.2 Pitch-excited LPC

Like all pitch-excited vocoders, the pitch-excited LPC is a *fully parametric* coder. This means that the coded speech is characterized entirely by the time-varying parameters of a speech synthesis model (see Figure 1.2). As discussed in the introduction, this synthesis model basically has two components: the excitation model and the vocal tract model. LPC techniques are used to parameterize the vocal tract model in this synthesizer. In all linear predictive coding techniques, the vocal tract is modeled as a linear time-varying filter. The parameters of the linear filter are obtained through a linear predictive analysis of the speech signal. In pitch-excited LPCs, the excitation signal is also fully parametric, and the parameters are extracted using a pitch detector. For other classes of LPCs, the excitation is represented and extracted in different ways.

Figures 5.1 and 5.2 show block diagrams of a complete pitch-excited LPC analyzer (transmitter), and a synthesizer (receiver). In the transmitter, the vocal tract model parameters and the excitation model parameters are extracted, quantized, coded, multiplexed, and transmitted. In the receiver, the coded parameters are extracted and used to synthesize the coded speech.

A pitch-excited LPC transmitter does two different types of analyses: the excitation analysis (pitch detector) and the vocal tract analysis (LPC analysis). In Figure 5.1, the pitch detector is shown at the top of the figure; it operates directly on the input signal, $s[n]$. The outputs of the pitch detector include a voicing decision (voiced or unvoiced) for each frame, and, for voiced frames, a pitch period. These parameters are coded and multiplexed into the output data stream. The primary control parameter of interest in this laboratory manual for the pitch detector is the frame interval between successive estimates of pitch and voicing.

The LPC analysis is shown in the bottom half of Figure 5.1. In the analysis section, the speech is first passed through a pre-emphasis filter. The purpose of this filter is to reduce the dynamic range of the speech spectra, which, in turn, improves the numerical properties of the LPC analysis algorithms. The pre-emphasized speech is then windowed into frames for analysis. The window type, window length, and window frame interval are important parameters of the LPC coder. After the window has been applied, a correlation analysis is performed on the resulting finite-length signals. The number of points used for the correlation analysis and the associated number of coefficients used for the LPC analysis are

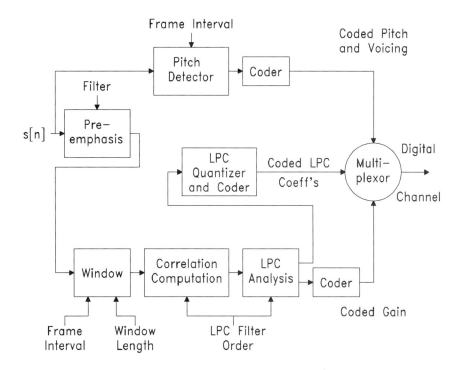

Figure 5.1. Block diagram of a pitch-excited LPC transmitter.

the primary control parameters for the correlator and subsequent LPC analyzer. The results of the LPC analyses for each frame are a gain parameter and a set of LPC filter parameters. Both of these are quantized, coded, and multiplexed into the output data stream for transmission or storage.

The block diagram of a receiver for the pitch-excited linear predictive vocoder is shown in Figure 5.2. The basic speech synthesizer consists of an excitation signal that is an input to a time-varying vocal tract filter. The excitation generator includes the pulse generator, the noise generator, the voicing switch, and the gain. The vocal tract filter is formed by the linear predictor operating in a recursive loop. The de-emphasis filter is the inverse filter for the pre-emphasis filter found in the transmitter.

Operation of the receiver can be summarized as follows. Data from the input digital channel is demultiplexed into its three components: pitch and voicing, gain, and LPC coefficients. The pitch data is used to control the pulse rate on the pulse generator while the voicing data is used to control the position of the voicing switch. The gain data is used to control the amplitude of the excitation signal, and hence the loudness of the output speech. The LPC coefficients are used to control the vocal tract filter. The role of the de-emphasis filter that follows the vocal tract

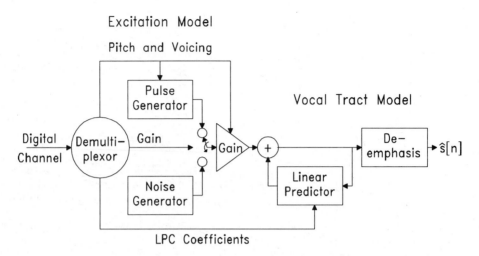

Figure 5.2. Block diagram of a pitch-excited LPC receiver.

filter is to reverse the spectral shaping imposed on the speech at the transmitter by the pre-emphasis filter.

In the LPC model, the synthesis filter is a representation of the acoustic filtering effect of the vocal tract. The synthesis filter is usually realized as an all-pole recursive digital filter whose input approximates the excitation to the vocal tract and whose output is the synthetic speech. As discussed in Chapter 1, high quality speech requires a bandwidth of 6 kHz or greater, while telephone speech normally requires a bandwidth of about 3.2 kHz with a sampling rate of 8000 samples per second (i.e., a sampling period of 125 microseconds). However, since the vocal tract is a mechanical system, its rate of change is much lower than this. These slowly changing characteristics of the speech signal are modeled by the (relatively) slowly changing parameters of the vocal tract filter. This results in a time-varying synthesis filter whose coefficients are changed slowly (and usually periodically) in time.

5.3 Vocal Tract Model

As shown in Figure 5.2, the speech synthesizer used by the LPC receiver can be divided into two parts: the excitation model and the vocal tract model. The vocal tract model has two components: the vocal tract filter and the de-emphasis filter. As will be discussed, the vocal tract filter can be implemented in several forms. In the simplest implementation, the vocal tract is modeled as a direct form IIR filter as shown in Figure 5.3. In all forms, the vocal tract filter is characterized by P parameters, where P is normally 10-12 for speech sampled at 8000 samples per second.

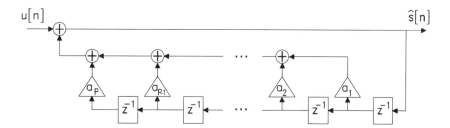

Figure 5.3. Direct form implementation of the vocal tract filter.

The primary task of the transmitter with regard to the vocal tract filter is to periodically analyze the input speech (usually 40–100 times per second) to estimate, quantize, code, and transmit the vocal tract parameters necessary to implement the vocal tract filter at the receiver. As shown in Figure 5.1, this is basically accomplished in four steps: the pre-emphasis filter, the correlation computation, the LPC analysis, and the LPC quantization and coding. Each of these steps will be discussed. However, because of their importance, the correlation computation and the LPC analysis will be discussed first.

5.3.1 Correlation Computation and the LPC Analysis

The LPC analysis operates on frames of data. The heart of the LPC is the linear predictor. In the linear predictive model, it is assumed that the speech signal is an autoregressive process that can be represented as

$$\hat{s}[n] = \sum_{i=1}^{P} a_i \hat{s}[n-i] + Gu[n] \tag{5.1}$$

where $\hat{s}[n]$ is the synthetic speech produced by the model; $u[n]$ is the excitation signal; a_i $i = 1, \ldots, P$ are the prediction parameters; and P is the order of the predictor. In this expression, G is the gain parameter that is used to match the energy of the synthetic speech to that of the original speech signal. In the z-transform domain, $\hat{S}(z)$ is the output of the filter, $H(z)$, to the input signal, $U(z)$. $H(z)$, the LPC synthesis filter, is given by

$$H(z) = \frac{1}{1 - A(z)} \tag{5.2}$$

where $A(z)$ is the prediction filter given by

$$A(z) = \sum_{k=1}^{P} a_k z^{-k}. \tag{5.3}$$

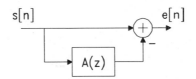

Figure 5.4. Block diagram of LPC inverse filtering.

In these terms, $\hat{S}(z)$ can be written as

$$\hat{S}(z) = H(z)U(z) = \frac{U(z)}{(1-A(z))} = \frac{U(z)}{(1-\sum_{k=1}^{P} a_k z^{-k})}. \qquad (5.4)$$

As shown in the pitch-excited LPC model in Figure 5.2, the excitation signal is assumed to be a pulse train for the voiced speech and white noise for the unvoiced speech. The period of the pulse train is equal to the pitch period of the speech signal. Thus the parameters of this synthesis model are the predictor coefficients (a_i's), the pitch period, the voiced/unvoiced parameter, and the gain parameter (G). The predictor coefficients are the parameters of the vocal tract, and the others are the excitation signal parameters.

In the LPC analysis of speech, the parameters of both the excitation model and the vocal tract model are approximated from the input speech signal. As can be seen from Equation 5.4, the transforms of the vocal tract filter transfer function and the excitation are multiplied together in the z-transform domain. From the frequency-domain point of view, it turns out that the vocal tract model carries the spectral envelope information, and the excitation model provides the information about the spectral detail of the speech.

Time-Varying Vocal Tract Model

In an LPC model, the vocal tract is represented by the all-pole filter $H(z)$. Because speech is a time-varying process, $H(z)$ must be a time-varying filter whose coefficients are changed in time. Because the vocal tract moves relatively slowly, speech can be assumed to be a random process whose properties vary slowly. This leads to the basic *short-time stationarity* assumption used for LPC analysis. This assumption states that the speech signal can be considered to be stationary during a window of L samples as long as L is small enough. This assumption leads to a modeling of speech by a succession of fixed filters, $H(z)$'s, whose coefficients stay constant within the window. The coefficients of $A(z)$, a_i $i = 1, \ldots, P$ are obtained through a linear predictive analysis of the speech signal.

There are a number of ways to look at linear predictive analysis. One of the most instructive is illustrated in Figure 5.4. From this perspective, the linear predictor, $A(z)$, generates an estimate of the speech signal, $\hat{s}(n)$, from the input speech

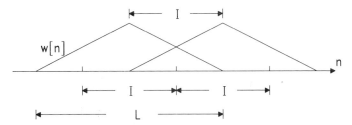

Figure 5.5. Sliding windows are applied to the speech signal for autocorrelation analysis. The window length, L, is independent of the frame interval, I.

signal, $s(n)$. This estimate is then subtracted from the original speech, giving the error signal, $e(n)$, which is called the *prediction residual signal*. The error signal is generated by the inverse filter given by

$$\frac{1}{H(z)} = 1 - A(z). \tag{5.5}$$

The predictor coefficients are estimated by minimizing the energy of the prediction residual, **E**, given by

$$\mathbf{E} = \sum_n e^2[n] \tag{5.6}$$

with respect to the prediction coefficients. In this expression, $e[n]$ is the output of the inverse filter and is given by

$$e[n] = s[n] - \sum_{i=1}^{P} a_i s[n-i]. \tag{5.7}$$

There are several approaches that can be used to obtain the predictor coefficients. Two of the most popular least-squares methods, the *autocorrelation* method and the *covariance* method, are discussed here. Both methods are primarily time-domain techniques that are easily implemented on DSP processors.

Autocorrelation Method

In the autocorrelation method, a moving window is used to divide the speech into frames. This process is illustrated in Figure 5.5. For each placement of the window, usually from 10 to 30 msec apart, the speech signal is windowed to create one analysis frame of the signal. The resulting signal is infinite in extent, but is zero everywhere outside the window. Thus it is possible to compute the true autocorrelation function for the entire signal. The ith analysis frame is given by

$$s_i[n] = s[n]w_i[n] \tag{5.8}$$

where $w_i[n]$ is the ith analysis window. The ith analysis window is normally given by

$$w_i[n] = w[n - iI] \tag{5.9}$$

where I is the analysis frame interval. The autocorrelation of the analysis frame is defined by

$$R[|k|] = \sum_{-\infty}^{+\infty} s_i[n]s_i[n + |k|]. \tag{5.10}$$

The window function, $w[n]$, is usually chosen to be a tapered function (e.g., a Hamming window) of size L where L is the analysis window size. The minimization of the mean residual energy leads to the matrix normal equation

$$\mathbf{Ra} = \mathbf{r} \tag{5.11}$$

where $\mathbf{a} = \{a_1, \ldots, a_P\}$ is the vector of LPC coefficients, and $\mathbf{R}$ is the matrix of autocorrelation coefficients and is defined by

$$\mathbf{R}[i,j] = R[|i-j|] = \sum_{n=-\infty}^{+\infty} s_i[n]s_i[n - j + i], \tag{5.12}$$

and $\mathbf{r} = \{R[1], \ldots, R[P]\}$. Matrix $\mathbf{R}$ is a symmetric Toeplitz matrix that can be solved efficiently using Durbin's algorithm [5]. This algorithm is recursive and uses the Toeplitz structure of $\mathbf{R}$ to solve for the LPC coefficients efficiently. This algorithm can be summarized by the following set of equations:

$$\mathbf{E}^0 = R[0] \tag{5.13}$$

$$k_i = \left[R[i] - \sum_{j=1}^{i-1} a_j^{i-1} R[i-j]\right] / \mathbf{E}^{i-1} \tag{5.14}$$

$$a_i^i = k_i \tag{5.15}$$

$$a_j^i = a_j^{i-1} + k_i a_{i-j}^{i-1}, \quad 1 \leq j \leq i-1 \tag{5.16}$$

$$\mathbf{E}^i = (1 - k_i^2)\mathbf{E}^{i-1}. \tag{5.17}$$

Equations 5.13 to 5.17 are solved recursively for $i = 1, \ldots, P$. The coefficients k_i for $i = 1, \ldots, P$ contain the same information as the LPC coefficients, and are called *reflection coefficients* or *partial correlation* coefficients (known as PARCORs). It is possible to implement the vocal tract filter directly in terms of the PARCOR coefficients, as shown in Figure 5.6. In solving for the predictor of order P, the recursion also produces all the predictors of order 1 through $P - 1$. The quantity $\mathbf{E}^i$ is the energy of the prediction error with the predictor of order i. Since $\mathbf{E}^i$ is a positive quantity, Equation 5.17 shows that all PARCOR coefficients have a magnitude of less than one [5]. That is,

$$-1 < k_i < 1. \tag{5.18}$$

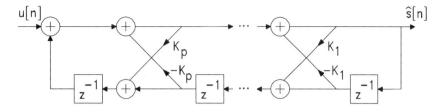

Figure 5.6. Lattice implementation of the vocal tract filter using PARCORs.

Because the LPC vocal tract filter is recursive, stability is an issue. It turns out that the condition of Equation 5.18 is also a necessary and sufficient condition for the stability of the vocal tract filter.

Covariance Method

In the covariance method, the speech signal is not windowed itself. Rather, the prediction error sequence $e[n]$ from Figure 5.4 is windowed and its energy is minimized. Thus the quantity defined by

$$\mathbf{E} = \sum_{n=-\infty}^{+\infty} e^2[n]w[n] \tag{5.19}$$

is minimized with respect to the prediction coefficients. This minimization results in a matrix equation of the form

$$\Phi \mathbf{a} = \phi \tag{5.20}$$

where $\mathbf{a}$ is the vector of predictor coefficients, the symmetric matrix Φ is defined as

$$\Phi[i,j] = \sum_{n=0}^{L-1} s[n-i]s[n-j], \tag{5.21}$$

and $\phi = \{\Phi[1,0], \ldots, \Phi[P,0]\}$. Since Φ is not a Toeplitz matrix, it cannot be solved as efficiently as the normal equations in the autocorrelation method, but relatively efficient solutions for solving symmetric equations do exist [5].

A Comparison of the Autocorrelation and Covariance Methods

Both the autocorrelation method and the covariance methods produce sets of LPC coefficients for the LPC vocal tract filter. Also, they both have comparatively efficient solutions. The primary difference is in the way the speech is windowed. In the autocorrelation method, the speech is first windowed, producing a short-time (and thus distorted) approximation to the original speech. Then an exact autocorrelation is computed, and optimal LPC coefficients are computed for this distorted speech segment. Mathematically, this is a cleanly tractable process, which is guaranteed to produce well-behaved (stable) vocal tract filters. Overall, the autocorrelation method has mostly been the preferred approach.

The covariance method, on the other hand, does not apply a window directly to the speech signal. Thus the speech is not distorted before processing begins. Rather, the covariance method windows the error, in effect windowing the period over which it is applied. This has the potential of giving better performance, because the speech has not been "pre-distorted," but the windowing of the error signal introduces another, less mathematically tractable form of distortion. It is possible for the covariance method to give unstable vocal tract filters.

Predictor Order

One of the decisions to be made in an LPC vocoder is the order of the LPC predictor. Because the residual energy decreases for every iteration of Durbin's recursion, the energy of the prediction error decreases as the number of the poles of the synthesis filter, P, increases. Since the final objective in a vocoder is to transmit the predictor coefficients to the receiver, and also because of the amount of computation, it is important to fix and limit the number of coefficients. One way of determining P is to find the threshold after which the error does not decrease significantly. If the threshold is t_e, and if

$$1 - \frac{E_{p+1}}{E_p} < t_e, \qquad (5.22)$$

then $P = p$ is a good choice. For speech, two poles (one pole pair) are used to model each formant. The speech signal also has spectral zeros, but since these have very little perceptual impact, they are not typically modeled as part of the vocal tract transfer function. In practice, for 8 kHz speech, predictor orders in the range of 10 to 16 are used.

5.3.2 Pre-emphasis

Voiced speech spectra normally have a drop-off of about 6-dB/octave,[1] which results in a high spectral dynamic range. In effect, the speech spectra are *tilted* into a slightly lowpass form. This high dynamic range usually results in an inaccurate approximation of the higher formants and sometimes results in ill-conditioned normal equations for the LPC analysis. In order to reduce this effect, the speech signal is often pre-emphasized prior to LPC analysis. This fixed pre-emphasis filter is usually of the form

$$V_{pre}(z) = 1 - \lambda z^{-1} \qquad (5.23)$$

where $V(z)$ is effectively a mild highpass filter with a single zero at λ. The constant, λ, controls the degree of pre-emphasis. Figure 5.7 shows the frequency response

---

[1] A 6-dB/octave drop-off means that the amplitude is scaled by $\frac{1}{2}$ for each frequency doubling.

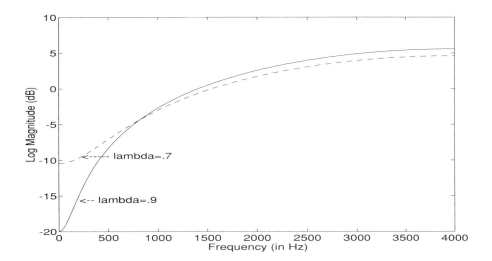

Figure 5.7. Frequency response of the pre-emphasis filter for $\lambda = 0.7$ and $\lambda = 0.9$.

of the pre-emphasis filter for $\lambda = 0.7$ and $\lambda = 0.9$. Although the optimal value of λ can be estimated statistically, the value is different for different talkers, and the analysis is not very sensitive to the value of λ.

In order to cancel the effects of the pre-emphasis, the fixed pre-emphasis filter is matched at the synthesizer by a fixed de-emphasis filter of the form

$$V_{de}(z) = \frac{1}{1 - \eta z^{-1}}. \tag{5.24}$$

Although λ and η are often chosen to be the same to cancel each other, different values for λ and η can result in better quality speech (e.g., $\lambda = 0.94$, $\eta = 0.74$) [7].

In simple terms, the pre-emphasis filter at the transmitter approximately removes the natural spectral tilt in voiced speech, while the de-emphasis adds it back in.

5.3.3 Window Considerations

A very important set of parameters for the linear predictive analysis are those concerning window operation. These include the type and size of the window used and the size of the analysis frame interval. Some typical windows are shown in Figure 5.8. When using the autocorrelation method, the window is repeatedly applied to the speech signal. To reduce the window's edge effects, tapered windows such as Hamming or Hanning windows are usually used. Such smooth windows produce better results than a rectangular window or other windows with sharp edges. The size of the window, L, is usually chosen to cover a few pitch periods for voiced

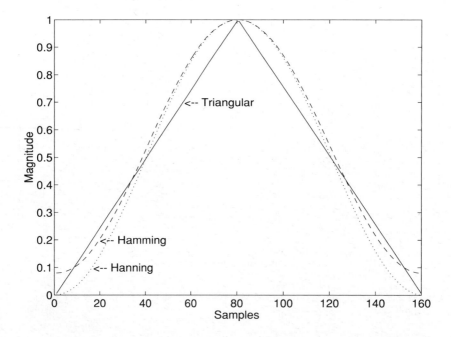

Figure 5.8. Some of the windows used in the LPC analysis. The figure shows the Hamming, Hanning, and triangular windows.

parameters	name	range	typical values
predictor order	P	1–20	10
LPC window length	L	160–320	240
LPC frame size	I	40–160	120
pre-emphasis factor	λ	0.7–0.95	0.8

Table 5.1. The parameters of the LPC vocal tract analysis.

speech (20-40 msec). This is needed to reduce the effects of the excitation signal on the estimation of the vocal tract filter parameters, and to obtain an accurate estimate of the speech spectrum. For such analysis frame sizes, autocorrelation and covariance methods produce similar results.

The analysis frame interval, I, determines the number of samples over which the resulting LPC coefficients will be used. The ratio I/L represents the amount of overlap between two adjacent analysis frames (see Figure 5.5). A 50% overlap is typically used in practice ($I = L/2$). Table 5.1 shows all the parameters of the LPC vocal tract analysis, their range, and some typical values.

EXERCISE 5.3.1. Vocal Tract Frame Rate

A very important parameter in a pitch-excited LPC vocoder is the vocal tract frame rate, or, equivalently, the vocal tract frame interval. Obviously, the bit-rate of the overall vocoder, which is equal to the vocal tract bit-rate plus the excitation bit-rate, is directly related to the vocal tract frame rate. Higher frame rates give better quality whereas lower frame rates give lower bit-rates.

The purpose of this exercise is to determine the effect of the vocal tract frame interval on the quality of the vocoder.

a. For this exercise, use the high female (*h_female.sig*) and the low male (*l_male.sig*) speech signals. For each, use the LPC Vocoder function (Signal Coding/Speech and Audio Coding/LPC/LPC Vocoder) to code the signals with a frame interval of 40, 80, 120, 200, and 300 samples. Use the standard parameters set for all other parameters. Listen to your results (File/Playback Signal).

The LPC vocoder function requires a pitch file. To operate most efficiently, use the precomputed pitch files, *h_female.pit* and *l_male.pit*, respectively. Otherwise, the required pitch files will need to be computed using the pitch detector (Signal Coding/Speech and Audio Coding/LPC/Pitch Detection).

b. In general, making the frame interval shorter makes the quality better. However, at some point, making the frame interval shorter does not continue to improve the quality because of the slowly varying nature of the vocal tract. This would be the minimum frame interval that would ever be required in an LPC vocoder. Making the frame interval shorter than this limit would only increase the bit-rate without increasing the quality. From your experiment in part a, what is your estimate of the minimum necessary frame interval? You may need to use headphones to hear the effect. Is it different for males and females?

c. As the frame interval increases, obviously both the quality and bit-rate will decrease. Based on your experiments in part a, what is the longest frame interval at which the coded speech is still intelligible? What is the longest frame interval for which the speech would still be usable?

d. Using the Waveform Compare function under the Signal Utils menu (Signal Utils/Waveform Compare), compare the coded and original speech signals. What differences do you see? Could you answer the questions from parts b and c from examining the waveforms?

EXERCISE 5.3.2. Window-Length Experiments

Window length has no effect on the bit-rate, so it can be set without bit-rate concerns. Basically, long windows give good frequency resolution and poor time resolution, and short windows give good time resolution and poor frequency resolution. Since speech coding requires good frequency resolution at certain times (in long voiced sequences) and good time resolution at other times (during transition periods and stop-consonant bursts), a fixed window length is always a compromise.

a. Starting with the standard parameter set and the high female ($h\_female.sig$) and low male ($l\_male.sig$) talkers, code several examples (Signal Coding/Speech and Audio Coding/LPC/LPC Vocoder) with different window lengths and the same fixed frame interval, and listen (File/Playback Signal) to the results. Your window lengths should include 5 msec (40 samples), 20 msec (160 samples), 30 msec (240 samples), 60 msec (480 samples), and 120 msec (960 samples).

b. From your results, what is the best window length? Is the analysis sensitive to the window length?

c. Using the Waveform Compare (Signal Utils/Waveform Compare) and Spectra Compare (Signal Utils/Compare Two Spectra) functions, compare the coded and original speech signals. What differences do you see? What time-domain and frequency-domain properties of the coded speech signal are related to the window length?

EXERCISE 5.3.3. Number-of-Coefficients Experiments

The number of LPC coefficients that are used in the vocal tract model affects both the quality and the bit-rate. From an acoustic modeling perspective, for an average-sized person, about ten LPC parameters should be needed to represent the (four) formants, the glottal pulse shaping, and the radiation damping, but this is only an approximation. In general, talkers with short vocal tracts (short people and children) require fewer LPC coefficients whereas talkers with long vocal tracts require more LPC coefficients.

The purpose of this exercise is to determine the number of LPC coefficients needed by a pitch-excited LPC vocoder.

a. Starting with the standard parameter set and the low male ($l\_male.sig$) and high female ($h\_female.sig$) talkers, code several sentences with different numbers of LPC coefficients (Signal Coding/Speech and Audio Coding/LPC/LPC Vocoder), and listen (File/Playback Signal) to the results. Your numbers should include 2, 6, 8, 10, 12, and 16.

b. In general, using more coefficients makes the quality better. However, at some point more coefficients no longer improve the effectiveness of the vocal tract model, and over-parameterization can even degrade the quality. The important question is, What is the maximum number of LPC coefficients that would ever be required in an LPC vocoder? From your experiment in part a, what is your estimate of the maximum number of required LPC coefficients? Is your estimate different for males and females?

c. As the number of coefficients decreases, obviously both the quality and bit-rate will decrease. Based on your experiments in part a, what is the smallest number of coefficients at which the coded speech is still intelligible?

d. Using the Waveform Compare (Signal Utils/Waveform Compare) and Spectra Compare (Signal Utils/Compare Two Spectra) functions, compare the coded and original speech signals for 2 and 10 coefficients. What differences do you see? Could you answer the questions from parts b and c from examining the waveforms? From examining the spectra?

e. Repeat the above experiments using the low male and high female talkers. How do your results change?

EXERCISE 5.3.4. **Pre-emphasis Experiments**

The purpose of the pre-emphasis filter is to whiten the input speech signal. The effect of pre-emphasis is to make the computation of the LPC filter coefficients a better-conditioned numerical process. If the input signal is not pre-emphasized, often high-frequency formants are not well resolved, and the vocal tract filter can even (in very rare cases) go unstable. A pre-emphasis filter, if not matched by a de-emphasis filter, changes the perceived character of the speech. The "distortion" is actually sometimes desirable, making the speech more intelligible in a noisy environment.

The pre-emphasis and de-emphasis filter coefficients are included in the LPC Vocoder (Signal Coding/Speech and Audio Coding/LPC/LPC Vocoder) functions. The coefficients, which should be equal if no change in quality is desired, typically have a range of 0.8-0.95. A value of 0.0 results in no pre-emphasis, and values greater than 1.0 result in unstable de-emphasis filters.

a. Using a pre-emphasis filter coefficient of 0.9 and a de-emphasis coefficient of 0.0 and 0.9, code the high female (*h_female.sig*) and low male (*l_male.sig*) speech signals using the LPC vocoder with the standard parameter set for the remaining parameters. Listen to your results. How do the coded speech signals with no de-emphasis compare to the coded signals with de-emphasis? How do the coded speech signals with no de-emphasis compare to the original signals?

b. Starting with the standard parameter set and the high female (*h_female.sig*) and low male (*l_male.sig*) speech signals, generate coded outputs using several different equal pre-emphasis/de-emphasis coefficients, including 0.0 (i.e., no pre-emphasis), 0.8, and 0.99. How much difference in quality can you hear?

c. Using the Waveform Compare (Signal Utils/Waveform Compare) functions, compare the coded speech signals. What differences do you see? What differences can you see from examining the waveforms?

d. Explain what happens when pre-emphasis is used without de-emphasis. Would this ever be useful?

5.4 Excitation Model

As noted earlier in this chapter, there are a number of popular types of linear predictive coders in use today. For the most part, all of these coders use a linear predictive vocal tract model, and most of them use similar LPC analysis techniques.

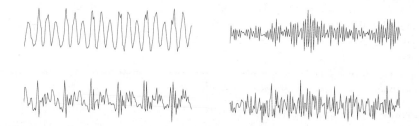

Figure 5.9. (*left*) A section of voiced speech and the corresponding residual signal shown below it, (*right*) A section of unvoiced speech and the corresponding residual signal shown below it.

The main difference between these coders is the way in which the input to the synthesis filter $H(z)$ is modeled and determined. In order to understand the nature of the excitation signal in an LPC environment, Equation 5.7 may be rewritten as

$$s[n] = \sum_{i=1}^{P} a_i s[n-i] + e[n]. \tag{5.25}$$

Comparing Equation 5.1 to Equation 5.25, it is obvious that if $Gu[n] = e[n]$, then the output of $H(z)$ will be equal to the original speech; that is, $\hat{s}[n] = s[n]$. Thus, in order for the LPC model to produce a good quality synthetic speech signal, $u[n]$ should be a good representation of the residual signal $e[n]$. Figure 5.9 shows two windowed sections of a speech signal and the corresponding residual signal for a predictor of order 10. As can be seen in this figure, the residual signal for voiced speech is a pseudo-periodic signal and for unvoiced speech is a noiselike signal. In pitch-excited LPC vocoders, the excitation signal is very simple and consists either of periodic impulses or white noise. Therefore, a simple model for the excitation signal, $u[n]$, is to have a simple periodic pulse train excitation for voiced speech and white noise excitation for unvoiced speech.

In order to produce such an excitation signal, two parameters should be obtained from the speech signal. First, the speech frame under analysis needs to be classified as voiced or unvoiced. Second, for the voiced segments, the pitch period needs to be determined.

EXERCISE 5.4.1. Voiced/Unvoiced Gain Experiments

The gain estimate in LPC analysis operates differently for voiced and unvoiced sounds. Thus the LPC pitch-excited vocoder in this text has a separate, fixed gain for voiced and unvoiced speech. In addition, a different ratio of voiced to unvoiced energy results in speech with different perceptual characteristics. High unvoiced gains result in speech that is perceived as "carefully articulated."

a. Starting with the standard parameter set and the high female ($h\_female.sig$) and the low male ($l\_male.sig$) speech signals, synthesize (Signal Coding/Speech and Audio Coding/LPC/LPC Synthesis) several output sentences with a voiced gain of 1.0 and different unvoiced gains, and listen to the results. Your unvoiced gains should include 0.0, 1.0, 5.0, 20.0, and several other values.

b. Use the waveform compare (Signal Utils/Signal Analysis) function to investigate the impact of the voiced/unvoiced ratio on the time-domain representation of the speech signal. How would you compare the ability of LPC analysis to model voiced and unvoiced sounds?

c. From your results, what is the best voiced/unvoiced ratio? Are there any conditions (e.g., in acoustic noise) where you might use a different gain ratio? Explain.

5.4.1 Pitch Detection

There are many approaches to determining the pitch period of voiced speech. These procedures can generally be divided into time-domain and frequency-domain approaches. In time-domain approaches, the speech signal is processed to estimate the pitch period. In frequency-domain approaches, the spectral information and harmonic structure of the speech signal are used to estimate the pitch period.

The complexity and accuracy of these approaches vary considerably among different algorithms. Simple algorithms, such as center clipping and peak picking on the speech waveform or the inverse filtered speech (prediction residual), are examples of time-domain pitch detection methods. Techniques based on the short-time autocorrelation of speech or preprocessed speech have been studied extensively [5, 7].

An in-depth discussion of pitch detection techniques is beyond the scope of this text. However, a relatively high-performance pitch detector is included in the laboratory software. The primary control parameter for the pitch detector is the interval between "pitch and voicing" estimates. Pitch estimates are often made at the frame rate of the LPC analysis, but this is not necessarily the case. The pitch detector provided with DSPLAB generates an estimate of the pitch every 10 msec (every 80 samples for 8 kHz sampled speech).

5.4.2 Gain Computation

The gain parameter in the LPC model is used to produce a synthetic speech signal that has the same energy as the original speech signal. This can be achieved by matching the energy of the LPC filter output for an impulse input (or a white noise input) to the energy of the original speech [5]. This results in the following relation

parameters	name	range	typical values
predictor order	P	1–20	10
LPC frame size	I	40–160	120
de-emphasis factor	η	0.7–0.95	0.8

Table 5.2. The parameters of the LPC synthesis.

between the gain, G, and the autocorrelation coefficients of the speech signal:

$$G = \left[R(0) - \sum_{k=1}^{P} a(k)R(k) \right]^{1/2}. \quad (5.26)$$

Table 5.2 shows the list of LPC excitation model and synthesis parameters.

5.5 Quantization of LPC Model Parameters

A major component of all LPC coders is the quantization and coding of the speech-synthesis model parameters. The parameters that are transmitted at each analysis interval are:

1. Predictor coefficients a_i: $i = 1, \ldots, P$
2. Pitch period
3. Gain
4. Voicing parameter

The pitch period, the gain, and the voicing parameters can usually be quantized and coded using scalar quantizers. The LPC predictor coefficients can be represented in many different forms, some of which are more suitable for quantization than others. Direct quantization of the predictor coefficients is usually avoided because of the high number of bits required for each coefficient (8 to 10 bits are required). This many bits are required because the predictor coefficients are very sensitive to quantization errors. This means that small differences can have a significant impact on the resulting synthesis filter. The equivalent forms that are less sensitive to quantization that have been proposed and used include:

1. Reflection coefficients, k_i's (PARCORs)
2. Line spectrum pairs (LSPs), which are defined to be roots of the polynomials $P(z)$ and $Q(z)$ given by

$$P(z) = (1 - A(z)) + z^{-(P+1)}(1 - A(z^{-1})) \quad (5.27)$$
$$Q(z) = (1 - A(z)) - z^{-(P+1)}(1 - A(z^{-1})) \quad (5.28)$$

3. First P samples of the impulse response of $H(z)$, $h[n]$

4. Log area ratios (LARs), which are defined to be

$$\text{LAR}_i = \log\left(\frac{1-k_i}{1+k_i}\right) \tag{5.29}$$

5. Autocorrelation coefficients, $R[i]$'s
6. Cepstrum coefficients of $h[n]$, which can be obtained from the recursion

$$\hat{h}[n] = a_n + \sum_{k=1}^{n-1}\left(\frac{k}{n}\right)\hat{h}[k]a_{n-k}. \tag{5.30}$$

5.6 Spectral Estimation Using LPC

Techniques based on linear predictive analysis have been widely applied for spectrum estimation of many different types of signals. For speech signals, the frequency response of the synthesis filter, $H(z)$, tends to follow the spectral envelope of the speech spectrum. This can be seen by expressing the mean-squared prediction error in frequency-domain. In fact, linear prediction can be formulated in the frequency-domain, which also generates the same normal equations shown in Equation 5.11.

Applying the z-transform to Equation 5.7 results in

$$E(z) = (1 - A(z))S(z) \tag{5.31}$$

where $E(z)$ is the z-transform of the prediction residual and $S(z)$ is the z-transform of the speech signal. Using Parseval's theorem, the mean-squared error can be expressed as

$$\mathbf{E} = \sum_n e^2[n] = \frac{1}{2\pi}\int_{-\pi}^{\pi}|E(e^{j\omega})|^2. \tag{5.32}$$

Combining Equations 5.31 and 5.32, $\mathbf{E}$ can be expressed as

$$\mathbf{E} = \frac{G^2}{2\pi}\int_{-\pi}^{\pi}\frac{|S(e^{j\omega})|^2}{|H(e^{j\omega})|^2}d\omega. \tag{5.33}$$

Therefore, minimizing $\mathbf{E}$ is equivalent to minimizing the integral of the ratio of the energy spectrum of the speech signal to the energy spectrum of the impulse response of the filter, $H(z)$. Equation 5.33 shows the way the signal spectrum is approximated by the all-pole model spectrum. Obviously, in minimizing $\mathbf{E}$, regions where the ratio of the power spectrums is greater than 1 contribute more to the total error than the regions in which this ratio is smaller than 1. Thus the LPC spectral error favors a good representation of the spectral peaks of the signal. That is why $|H(e^{j\omega})|^2$ usually follows the spectral envelope of $|S(e^{j\omega})|^2$. Figure 5.10 shows an example of the FFT spectrum of a signal, and a 20-pole LPC spectral estimate of that signal. In Figure 5.10, it is possible to see the way in which the LPC filter behaves as an envelope over the harmonic structure of the excitation signal.

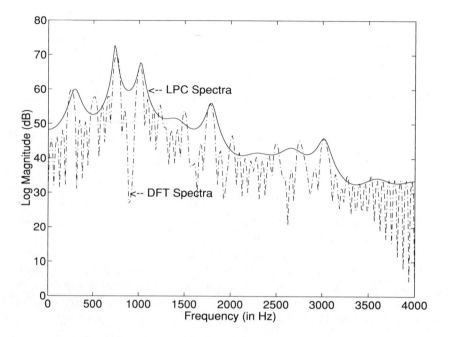

Figure 5.10. The FFT spectra and the 20-pole LPC spectra of a voiced segment of speech signal.

5.7 Exercises

EXERCISE 5.7.1. Optimal Pre-emphasis Coefficient

LPC analysis programs work best when the signals on which they are operating are approximately white. In Chapter 4, it was shown that the optimal predictor (which produces the maximum coding gain) is a filter that whitens the signal. Thus it is possible to find the optimal pre-emphasis coefficients using the long-term predictor analysis programs used in Chapter 4.

a. Using the predictor analysis program (Signal Coding/Speech and Audio Coding/LPC/Optimum Fixed Predictor), find an optimal pre-emphasis coefficient for five different sentences. How do the coefficients compare?

b. Using the LPC Vocoder (Signal Coding/Speech and Audio Coding/LPC/LPC Vocoder) function, code two of the sentences from part a using their optimal coefficients, a coefficient of 0.9, and a coefficient of 0.0. How much difference can you hear?

c. Using the Waveform Compare (Signal Utils/Waveform Compare) function, compare the coded and original speech signals. What differences do you see?

d. Based on the results of parts a, b, and c, characterize the importance of using an optimal pre-emphasis coefficient.

EXERCISE 5.7.2. **Pitch Modification**

Because the LPC receiver is a true speech synthesizer, the LPC analyzer is required to "take the speech apart" so it can be reassembled at the receiver. Thus the LPC representation (vocal tract and excitation) must be a full parametric representation, and it can be manipulated in its parametric form. One interesting type of manipulation changes the pitch signal to modify the character of the synthetic speech. In this exercise, you will modify the pitch signal in a number of ways to synthesize whispered speech, monotone speech, and pitch-modified speech.

a. Whispered speech is simply speech that has only a noise excitation. Synthetic whispered speech can be achieved by creating a pitch-period file of all zeros (unvoiced).

- Create an integer pitch-period file of all zeros. This can be done by creating an ASCII file of all zeros (Edit/Edit File(s)) and then converting it to a signal file (Edit/Convert Signal File) with an Output Range of 1.

- Code two sentences of your choice using the LPC Vocoder (Signal Coding/Speech and Audio Coding/LPC/LPC Vocoder) with a zero pitch file.

- What is the quality of the synthetic speech? Is it intelligible? Can you recognize the talker? Could you build a good LPC vocoder that does not use a pitch detector by using whispered speech?

b. Monotone, all-voiced speech can be achieved by supplying a pitch-period file with a constant value. This can be done by creating an ASCII file with a constant value (Edit/Edit File(s)) and then converting it to a signal file (Edit/Convert Signal File) with an Output Range of 1. The constant value is the pitch period that is in the form of a fixed number of samples at the sampling rate of the input signal.

- Create several integer pitch-period files with constant values. The values should include basic pitches of 50 Hz, 100 Hz, and 250 Hz.

- Code two sentences of your choice using the LPC Vocoder (Signal Coding/Speech and Audio Coding/LPC/LPC Vocoder) with a constant pitch file.

- What is the quality of the synthetic speech? Is it intelligible? Can you recognize the talker? Could you build a good LPC vocoder that does not use a pitch detector based on this technique? Have you ever been to the Atlanta airport?

c. The pitch-period files can also be modified using the signal arithmetic functions.

- Use the signal arithmetic functions to create pitch period files that are longer (multiply by a constant greater than 1), shorter (multiply by a constant less than 1), rising (multiply by a rising ramp), and falling (multiply by a falling ramp).
- Code two sentences of your choice using the LPC Vocoder (Signal Coding/Speech and Audio Coding/LPC/LPC Vocoder) with a modified pitch file.
- What is the quality of the synthetic speech? Is it intelligible? Can you recognize the talker? Could this technique be used to modify the apparent sex of the talker without losing intelligibility?

d. Use the Signal Clipping (Signal Utils/Signal Clipping) function to threshold the pitch file *h_female.pit*. Set the lower limit to 0 and the upper limit to 25. Use the output file as the pitch file for an LPC synthesis with an LPC coefficient file that was created from *h_female.sig*. Compare the synthesized speech with the synthesized speech from part b using the same input signal and a constant pitch of 25. Does the synthesized speech differ from the result in part b? Explain.

EXERCISE 5.7.3. Rate Change

An interesting use of a pitch-excited vocoder is to modify the time scale (speeding up or slowing down) of the synthetic speech without modifying the pitch. This can be achieved by analyzing the speech at one frame interval and synthesizing it at another. For example, if the speech is analyzed with a frame interval of 10 msec and synthesized with a frame interval of 20 msec, the synthetic speech will be at half speed.

a. Analyze (Signal Coding/Speech and Audio Coding/LPC/Simple LPC Analysis) two sentences with a 15-msec frame interval.

b. Synthesize (Signal Coding/Speech and Audio Coding/LPC/LPC Synthesis (pitch excited)) all the sentences at a frame increment of 80, 120, 160, and 240 and listen to the results.

c. At what rates are the sentences intelligible? At what rates are the talkers recognizable?

d. Plot the four waveforms together and comment on the comparison.

Waveform Coding with Adaptive Prediction

6.1 Introduction

Chapters 3 and 4 discussed a number of direct waveform coding techniques for encoding speech signals. On the whole, these methods used relatively few of the known characteristics of the speech signal and of aural perception in the encoding process. The resulting coding systems can be very high quality and are robust in acoustic noise, but only at the cost of quite a high bit-rate. Chapter 5, on the other hand, discussed the family of fully parametric pitch-excited vocoders. These coders are essentially speech synthesizers that incorporate a complete vocal tract model into the receiver. The resulting coders operate at low bit-rates with a high degree of intelligibility, but they can never be really high quality, and they operate well only on single-talker speech signals.

This chapter discusses the class of differential coders with adaptive predictors. Such coders can be viewed in two ways. First, they can be viewed as adaptive DPCM systems that use an adaptive linear predictor to track the short-term stationary statistics of a speech signal and achieve a high coding gain by better prediction of the speech waveform. Second, they can be viewed as *waveform-excited* vocoders such as the one shown in Figure 1.3. As discussed in Chapter 1, these systems can be considered to be like pitch-excited vocoders in which the excitation generator has been replaced with a coded waveform. Such systems, which have traditionally been called *adaptive-predictive coders* (APCs), can operate with higher quality than pitch-excited coders at a lower bit-rate than the other waveform coders. APCs are important because they are a vital link between waveform coding and parametric coding.

As discussed in Chapter 4, differential coders attempt to exploit the short-term predictability of a speech signal. Because of the nonstationarity of speech signals, differential PCMs with fixed predictors can achieve only a limited prediction gain. Adaptive prediction and adaptive quantization can be combined to produce a more sophisticated class of speech waveform coders. Adaptive-predictive coders are an important member of this class. These coders use linear prediction theory, as discussed in Chapter 5, to increase the prediction gain. In APC, predictor parameters are obtained either from the input speech or from the reconstructed output speech. Speech coders that obtain predictor parameters from the input speech must transmit the LPC parameters to the receiver, and are called *feed-forward* coders. Coders that obtain predictor parameters from the output speech do not need to transmit the predictor parameters because the output speech is available at the receiver. Such systems are called *feedback* coders. Feed-forward structures are usually more effective and are more robust for channel errors, but they require transmission of the side information necessary to control the adaptive predictor. On the other hand, feedback systems do not require transmission of any side information, but in general they operate with lower quality and are more susceptible to bit errors.

Another member of the class of waveform coders with adaptive prediction is the *adaptive-predictive coder with pitch prediction* (APC-PP). Adaptive-predictive coders with pitch prediction are DPCM-based coders that exploit the pitch redundancy of the speech signal as well as the short-term correlation of speech samples. Thus APC-PP is based on two adaptive predictors–a short-term predictor (STP) for spectral envelope estimation, and a long-term predictor (LTP) for pitch prediction. The addition of the LTP significantly improves the performance of the APC.

Noise-feedback coders are another member of the class of adaptive-predictive coders. In noise-feedback coding, the masking properties of the human auditory system are used to minimize the amount of audible noise in the reconstructed speech signal. To achieve this noise masking, the spectrum of the coding noise is shaped to be similar to the speech spectrum. In such systems, the energy in the noise spectrum will normally be less than the energy in the speech spectrum and the noise will be masked.

6.2 Adaptive-Predictive Coding

Figure 6.1 shows the block diagram of an APC. In this figure, the quantizer parameters are assumed to be fixed. In APC with adaptive feed-forward prediction (APC-APF), the predictor coefficients are obtained from the speech signal and are transmitted as side information. In contrast, APC with adaptive feedback prediction (APC-APB) uses the coded signal to determine the predictor parameters. In Figure 6.1, the bold solid lines show the feed-forward path and the dashed lines show the feedback path for the adaptive predictor.

In a feed-forward strategy, the predictor parameters are determined by a linear predictive analysis using the methods described in Chapter 5. The parameters that

Sec. 6.2 Adaptive-Predictive Coding

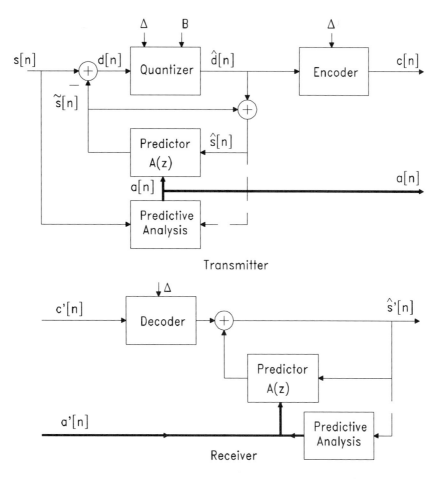

Figure 6.1. Block diagram of adaptive-predictive coding (APC).

must be extracted and coded for the APC are exactly the same as those for the vocal tract filter in a pitch-excited LPC. The disadvantages of the feed-forward system include the coding delay that is required for predictor estimation (as in LPC), and the extra bits needed to transmit the side information. The advantage of the feed-forward system is that it is less sensitive to transmission bit errors. As in the pitch-excited LPC, the predictor parameters are updated on a frame-by-frame basis to minimize the amount of side information.

In the feedback form of an APC system, the predictor is obtained directly from the coded speech samples by also using linear predictive analysis. Because a speech sample must be available in order to be used, only the past values of the speech signal are utilized to estimate the predictor coefficients. Also, because there is no side information required, the predictor can be updated at any rate without affecting the bit-rate. Thus the predictor coefficients are updated at a high rate,

parameters	name	range	typical values
predictor order	P	1–10	1
LPC frame size	I	80–360	160
quantizer max.	S_{max}	1–32767	32767
number of bits	B	1–16	5

Table 6.1. The parameters of APC with feed-forward prediction and uniform quantizer.

parameters	name	range	typical values
predictor order	P	1–10	4
LPC frame size	I	80–360	160
adaptive quantizer parameters (see quantizer parameter tables)	–	–	–

Table 6.2. The parameters of APC with feedback adaptive quantizer. The quantizer parameters are shown in Chapter 3.

sometimes at every sample. This adds to the computational complexity of the feedback system. A major disadvantage of the feedback system is that with coarse quantization, the predictor is often not very good, which can seriously affect the system performance at low bit-rates. In the feedback mode, the predictor parameters are often obtained using recursive analysis techniques such as the *recursive autocorrelation method* [10], the *least mean square* (LMS) method, or the method of *steepest descent* [6]. In this text, we emphasize feed-forward systems because of their robustness and ease of implementation. The parameters of an APC-APF (which will be referred to as APC from now on) with a fixed uniform quantizer are shown in Table 6.1. The predictor parameters are obtained through an LPC analysis of the speech signal. In addition to the predictor order, P, and the LPC frame size, I, the quantizer parameters must be specified.

In this text, APC is considered to be a DPCM with both an adaptive predictor and an adaptive quantizer. In APC, the quantizer can either be feed-forward or feedback adaptive. Figures 6.2 and 6.3 show the block diagrams of the APC system with feed-forward and feedback adaptive quantizers. In the feed-forward scheme, the quantizer parameters are transmitted as side information. Table 6.2 shows the parameters of the APC with feedback adaptive quantization.

EXERCISE 6.2.1. APC

The purpose of this exercise is to understand the effect of adaptive prediction in the performance of waveform coders. Two parameters are of particular importance: the predictor order, P, and the frame interval, I. Higher predictor orders and shorter frame sizes will generally produce better performance (higher SNR) but also higher bit-rates.

a. Assuming a step size quantizer with 2, 3, 4, 5, and 6 bits, and using the high female (*h_female.sig*) and low male (*l_male.sig*) talkers, analyze the sentences (Signal Coding/Speech and Audio Coding/APC/Optimum Adaptive Predictor) and run APC (Signal Coding/Speech and Audio Coding/APC/APC Step Size Adapt. Quant.) from the APC menu using the standard parameters. Sketch the SNR versus bit-rate diagram. Compare APC with step size adaptive quantizer to the ADPCM (Signal Coding/Speech and Audio Coding/Waveform/ADPCM Step Size Adapt. Quant.) with the optimum fixed predictor (Signal Coding/Speech and Audio Coding/Waveform/Optimum Fixed Predictor) of the same order. Recall that the ADPCM results are available from the exercises in Chapter 4.

b. For a fixed frame size (e.g., 120), a fixed number of bits (e.g., 6), and a sentence of your choice, find the predictor order above which the coding gain improvements are not significant.

c. For a fixed predictor order (e.g., 10), find the smallest LPC frame size below which no significant coding gain (Signal Coding/Speech and Audio Coding/APC/SNR S/N Ratio) improvements are observed. You should use LPC frame sizes 80 and 500 in your tests. Listen to your results and comment on the differences in the effects of LPC frame size between LPC pitch-excited vocoders and APCs.

6.3 APC with Pitch Prediction (APC-PP)

Another way to predict the speech signal and increase the coding gain is to use *pitch redundancy* for voiced sounds. During voiced speech, the waveform is almost periodic, and this *pseudo-periodicity* can be used to help predict the speech waveform. This is normally accomplished using a *long-term predictor* (LTP). Like the short-term predictor, the LTP is also a linear predictor, but whereas the STP prediction is based on adjacent samples, the LTP prediction is based on samples from one or more pitch periods in the past. Long-term predictors can be used to remove the pitch redundancy of the speech signal. Just as the STP produces an estimate of the spectral envelope of the speech signal, the LTP represents the harmonic structure of the speech spectrum. Adaptive-predictive coders with pitch prediction attempt to exploit both short-term and long-term redundancies of the speech signal to produce an effective waveform coder at moderate bit-rates.

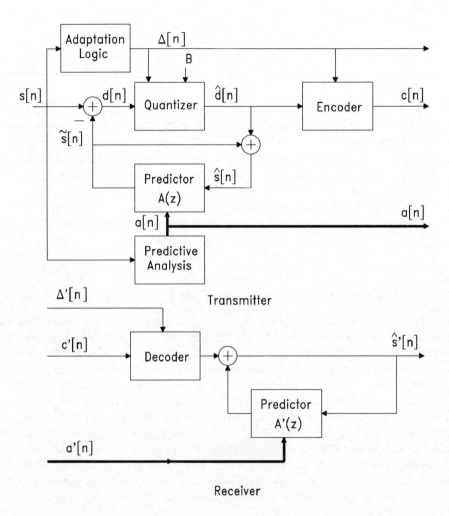

Figure 6.2. Block diagram of APC with a feed-forward adaptive quantizer.

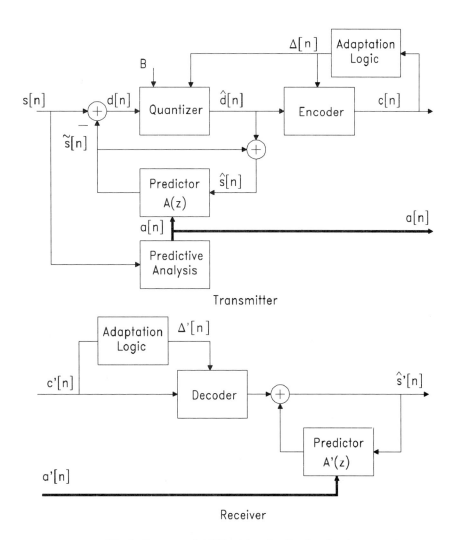

Figure 6.3. Block diagram of APC with a feedback adaptive quantizer.

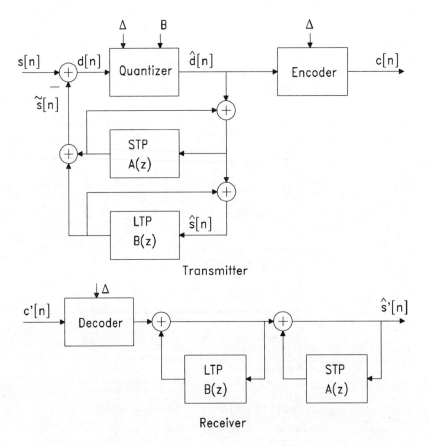

Figure 6.4. Block diagram of pitch-predictive APC (APC-PP).

Figure 6.4 shows the block diagram of an APC-PP. The STP is a prediction filter of order P given as

$$A(z) = \sum_{k=1}^{P} a_k z^{-k} \qquad (6.1)$$

and the LTP is usually of the form

$$B(z) = \sum_{k=-\ell}^{\ell} \beta_k z^{-\gamma-k} \qquad (6.2)$$

where β's are LTP coefficients and $2 \times \ell + 1$ is the order of the predictor. γ is the long-term delay that usually corresponds to a pitch period or an integer multiple of the pitch period. Because the corresponding synthesis filter may become unstable, a single-tap LTP of the form

$$B(z) = \beta z^{-\gamma} \qquad (6.3)$$

Sec. 6.3 APC with Pitch Prediction (APC-PP)

parameters	name	range	typical values
predictor order	P	1–10	1
LPC frame size	I	80–360	160
LTP frame size	N	10–80	40
long-term memory size	L	128–512	256
number of bits	B	1-5	3
quantizer parameters	–	–	–

Table 6.3. The parameters of pitch-predictive APC (APC-PP).

where $\beta < 1.0$ is usually preferred. At time n, the long-term delay is obtained by finding the value of γ that maximizes the normalized speech correlation given by

$$R_n[\gamma] = \frac{\sum_{n=0}^{N-1} s[n]s[n-\gamma]}{\sum_{n=0}^{N-1} s^2[n] \sum_{n=0}^{N-1} s^2[n-\gamma]} \quad (6.4)$$

where N is the long-term analysis block size. The LTP delay, γ, is usually chosen to be smaller than L, where L is the LTP memory size. The LTP gain, β, is given by the ratio of the speech autocorrelation at lags γ and 0,

$$\beta = \frac{R[\gamma]}{R[0]} = \frac{\sum_{n=0}^{N-1} s[n]s[n-\gamma]}{\sum_{n=0}^{N-1} s^2[n]}. \quad (6.5)$$

The above procedure for obtaining the LTP parameters is called the *open-loop* method. The *closed-loop* method, which is based on analysis-by-synthesis of speech, is discussed in Section 7.5. As will be illustrated more completely in Chapter 7, long-term predictors are very effective for improving the performance of differential coders.

Although the order in which the predictors are applied is not critical, the STP is usually applied before the LTP. The gain in the SNR provided by the combination of the two predictors is generally less than the sum of the individual gains. As shown in Figure 6.4, inverses of the long-term whitening filter, $1 - B(z)$, and the short-term whitening filter, $1 - A(z)$, are used to generate the coded speech signal at the receiver. The parameters of APC-PP consist of the STP and LTP parameters, and the quantizer parameters that are listed in Table 6.3.

EXERCISE 6.3.1. APC-PP

This exercise shows the effect of pitch prediction in APC. There are two possible ways of implementing the LTP in APC-PP. In the first procedure, LTP analysis is performed either on the speech signal or on the LPC residual signal. This is a feed-forward strategy where the LTP parameters are transmitted. In the second procedure, the LTP parameters are obtained from the coded speech signal in a feedback strategy. At low bit-rates, the feedback system will not be as effective as the feed-forward system, although feedback systems work very well in moderate bit-rate systems.

a. To run APC-PP with feed-forward LTP, generate the LTP parameters for the high female (*h_female.sig*) and low male (*l_male.sig*) talkers. You can do this by running the LPC Residual program (Signal Coding/Speech and Audio Coding/LPC/LPC Residual) which generates both the residual signal and the LPC coefficients. Run the LTP analysis (Signal Coding/Speech and Audio Coding/LPC/LTP Analysis) on this residual signal with the standard LTP analysis parameters. Display the LPC residual and LTP residual signals. The pitch periodicity of the LPC residual will be dramatically reduced in the LTP residual.

b. Listen to the LPC residual and LTP residual signals. Describe your results.

c. Assuming a step size adaptive quantizer with 2, 3, 4, 5, and 6 bits, and using the high female (*h_female.sig*) and low male (*l_male.sig*) talkers, run APC-PP (Signal Coding/Speech and Audio Coding/APC/APC-PP Step Size Adapt. Quant.) with the standard parameters. Sketch the SNR and SEGSNR, versus bit-rate diagram. Compare APC-PP with APC with no pitch predictor in terms of SNR, SEGSNR and speech quality.

6.4 Noise-Feedback Coding

Noise-feedback coding (NFC) refers to a modified version of the DPCM coder that attempts to exploit the masking properties of the human auditory system. In NFC, the subjective loudness of the coding noise is reduced by shaping the coding noise spectrum to match the spectrum of the input speech signal. A general DPCM coder configuration such as the one discussed in Figure 4.1 yields approximately white coding noise (assuming fine quantization). This noise would not be audible in regions of the speech spectrum that contain substantial signal energy, but at frequencies where there is little or no speech energy, it would be quite audible. In order to reduce the audible noise, the noise spectrum must be shaped to match the speech spectrum. This results in a greater total noise energy, but because of the masking of the coding noise by the speech, less perceived noise.

The DPCM system of Figure 4.1 can be equivalently represented as shown in Figure 6.5. This alternate representation of DPCM can be used to obtain waveform coders with noise shaping. Because the quantizer input in DPCM, $d[n]$, can be expressed as

$$d[n] = s[n] - \tilde{s}[n] = s[n] - \sum_{i=1}^{P} a_i \hat{s}[n-i], \qquad (6.6)$$

and because

$$\hat{s}[n] = s[n] + q[n] \qquad (6.7)$$

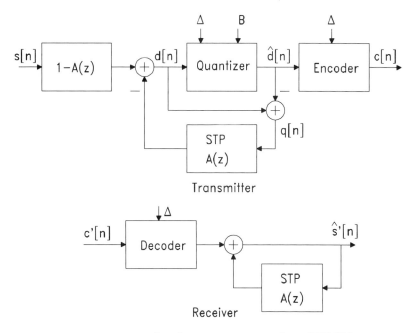

Figure 6.5. An alternate representation of DPCM.

where $q[n]$ is the coding noise, $d[n]$ can be expressed as

$$d[n] = s[n] - \sum_{i=1}^{P} a_i s[n-i] - \sum_{i=1}^{P} a_i q[n-i]. \quad (6.8)$$

Equation 6.8 can be expressed in z-transform domain as

$$D(z) = S(z)(1 - A(z)) - Q(z)A(z). \quad (6.9)$$

Figure 6.5 shows the alternate implementation of DPCM based on Equation 6.9. As seen in Equation 6.9, $d[n]$ has two components. The first component is a version of the speech signal that has been filtered by the short-term whitening filter, $1 - A(z)$. The second component is the quantization noise that has been filtered by the prediction filter, $A(z)$. As shown in Figure 6.5, the speech component can be obtained in a feed-forward path and the noise component can be obtained in a feedback path. Assuming these two components are independent, the feedback filter, $A(z)$, can be replaced by a general feedback function, $F(z)$. In such a system, the input speech signal and the quantization noise are processed with separate filters. This system is shown in Figure 6.6. It can be shown [6] that in Figure 6.6, the z-transform of the output coding error signal

$$y[n] = s[n] - \hat{s}[n] \quad (6.10)$$

118   Waveform Coding with Adaptive Prediction

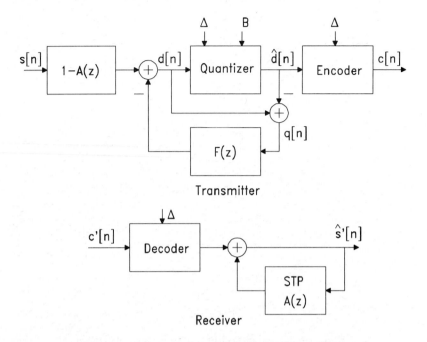

Figure 6.6. Block diagram of noise feedback coding.

is given as

$$Y(z) = Q(z)\frac{1-F(z)}{1-A(z)} = \frac{Q(z)}{W(z)} \tag{6.11}$$

which is a version of the quantization noise, $q[n]$, that has been filtered by the inverse of the noise shaping filter, $W(z)$. Thus by changing $F(z)$ and hence $W(z)$, the spectrum of the coding noise can be shaped in an arbitrary manner.

Open-Loop DPCM (D*PCM) DPCM is a special case of the system of Figure 6.6 with $F(z) = A(z)$. Another special case of this system is when $F(z) = 0$, which results in no feedback loop in the coder. This system is referred to as *open-loop* DPCM or D*PCM (pronounced D *star* PCM). As shown in Figure 6.7, D*PCM consists of a pre-filter, a quantizer, and a post-filter. Since the post-filter,

$$H(z) = \frac{1}{1-A(z)}, \tag{6.12}$$

represents the speech spectral envelope, the post-filter shapes the quantization noise spectrum to be identical to the input speech spectrum. Because the quantization noise, $q[n]$, is filtered through $H(z)$ (an IIR filter whose gain is larger than 1), D*PCM always has a lower SNR than DPCM [6]. There is, however, a risk of error accumulation in the synthesizer that can degrade the performance. Thus a compromise between D*PCM and DPCM is usually preferred.

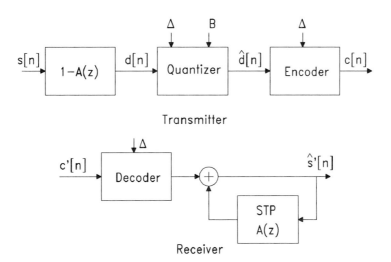

Figure 6.7. Block diagram of the open-loop DPCM (D*PCM).

DPCM with Noise Feedback Because the quantization noise, $q[n]$, is filtered by the inverse of the error-shaping filter, $W(z)$, and the speech is passed through $H(z)$, the spectrum of the noise component of the output can be shaped arbitrarily without affecting the speech component of the output. A compromise choice for $F(z)$ between the two extremes ($F(z) = A(z)$ for DPCM, and $F(z) = 0$ for D*PCM) is $A(\theta z)$, where $0 < \theta < 1$. Extreme values of $\theta = 1$ and $\theta = 0$ yield DPCM and D*PCM respectively. Typical values of θ for proper noise feedback are in the range 0.7–0.95. Such a choice for θ increases the bandwidths of the zeros of $1 - F(z)$, causing the noise to peak up in formant regions where it is less audible. Figure 6.8 shows the frequency response of the inverse error-weighting filter, $(W(z))^{-1}$, for a frame of voiced speech with $\theta = 0$, $\theta = 0.8$, and $\theta = 1$. The inverse filter emphasizes the coding noise where the signal has large energy ($W(z)$ is small, consequently making the ratio $Q(z)/W(z)$ large) and de-emphasizes the noise where the signal has small energy. In practical applications, a peak-limiter, with a peak parameter C, is applied in the feedback loop to ensure stability, especially with coarse quantization. Although NFC produces an SNR that is between that of DPCM and D*PCM, its subjective performance is better than both of them, especially at low bit-rates.

APC with Noise Feedback and Pitch Prediction Adaptive-predictive coding can be combined with noise feedback to improve the performance of APC. A block diagram of an APC-PP with noise feedback (APC-PPNF) is shown in Figure 6.9. APC-PPNF has an extra long-term predictor loop compared to the NFC of Figure 6.6. Adaptive spectrum prediction, adaptive pitch prediction, and adaptive noise shaping result in an effective coder at medium bit-rates (9.6–24 kbps).

Table 6.4 shows the parameters of the APC-PPNF. The parameters consist of

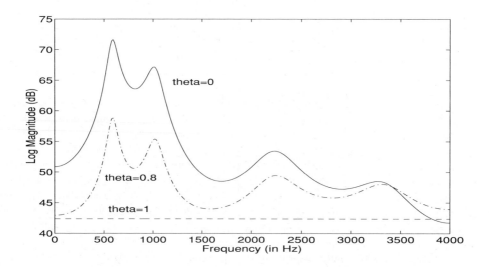

Figure 6.8. Frequency response of the inverse error-weighting filter, $(W(z))^{-1}$, for $\theta = 0$, $\theta = 0.8$, and $\theta = 1$.

the predictor parameters; the noise feedback parameter, θ; the peak-limiter parameter, C; and the quantizer parameters. DPCM, D*PCM, and NFC can be considered to be special cases of APC-NFPP. DPCM and D*PCM can be implemented by having $\theta = 1$ and $\theta = 0$ and the long-term predictor coefficient $\beta = 0$. NFC is obtained with $0 < \theta < 1$ and $\beta = 0$, and APC-PPNF is obtained when $\beta \neq 0$ and $0 < \theta < 1$. Table 6.5 summarizes these cases.

EXERCISE 6.4.1. APC with Noise Feedback

The purpose of this exercise is to show the effect of noise feedback on the quality of APC systems. The number of coders that can be obtained by using different STP, LTP, noise feedback, and quantization strategies is very large. DSPLAB directly provides some of the most important ones for study. They include D*PCM, APC-NF, and APC-PPNF with different quantizers.

a. Assuming a step size adaptive quantizer with 2, 3, 4, 5, and 6 bits, and using the high female ($h\_female.sig$) and low male ($l\_male.sig$) talkers, analyze the sentences (Signal Coding/Speech and Audio Coding/APC/Optimum Adaptive Predictor) and run APC-NF (Signal Coding/Speech and Audio Coding/APC/APC-NF Step Size Adapt. Quant.) from the APC menu using the standard parameters. Sketch the SNR versus bit-rate diagram. Compare APC-NF to APC in terms of SNR, SEGSNR, and speech quality.

b. Set the Error Weighting, $\theta = 0$, to obtain a D*PCM at 2, 3, 4, 5, and 6 bits/sample (Signal Coding/Speech and Audio Coding/APC/APC-NF Step Size

Sec. 6.4 Noise-Feedback Coding

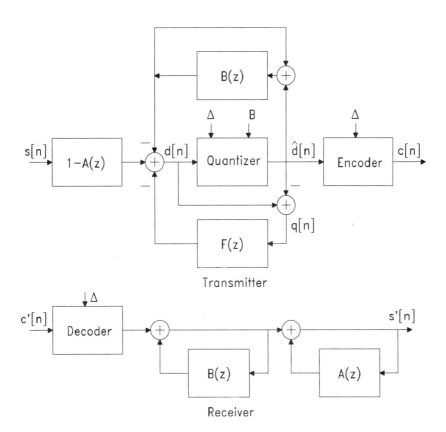

Figure 6.9. Block diagram of an APC-NF

parameters	name	range	typical values
predictor order	P	1–10	10
LPC frame size	I	80–360	160
LTP frame size	N	10–80	40
long-term memory size	L	128–512	256
error-weighting factor	θ	0–1	0.8
feedback-clipping limit	C	500–10000	3000
number of bits	B	1–4	3
quantizer parameters	–	–	–

Table 6.4. The parameters of APC with long-term (pitch) prediction and noise feedback (APC-PPNF).

coder type	θ	β
DPCM, ADPCM	1	0
D*PCM	0	0
NFC	0–1	non-zero
APC-PPNF	0–1	non-zero

Table 6.5. The specific values of θ and β in APC-NFPP to obtain DPCM, D*PCM, and NFC.

Adapt. Quant.). Compute the SNR (Signal Coding/Speech and Audio Coding/APC/SNR S/N Ratio) and compare to the SNR values obtained in part a. How does D*PCM compare to APC-NF in quality?

c. With 2, 3, 4, 5, and 6 bits/sample, run APC-PPNF (Signal Coding/Speech and Audio Coding/APC/APC-PPNF Step Size Adapt. Quant.) and feed-forward LTP using the standard parameters. Compare the performance of this coder to that of APC without noise feedback, and APC without LTP, APC-NF.

6.5 Residual-excited LPC

A basic problem with all forms of APC is that the excitation signal that must be coded contains as many samples as the original speech signal. Thus, even if only a very few levels were used to code the excitation signal, the bit-rate for the excitation is much higher than the bit-rate for the linear predictor, and is certainly much higher than the bit-rate required for the excitation signal in a pitch-excited LPC vocoder.

Residual-excited LPC (RELP) is a waveform-excited linear predictive coder in which the LPC synthesis filter is excited by a waveform that is directly derived from the LPC residual signal, but which has fewer samples than the actual residual. Such an approach has the advantage that it does not require any voiced/unvoiced classification of speech, or any pitch detection. Instead, a baseband filtered and downsampled version of the residual signal is coded and transmitted. This signal still contains a great deal of information (pitch, energy, etc.) from the original residual, and is used to generate a new, full-band excitation signal at the receiver.

Figure 6.10 shows the block diagram of a RELP vocoder. The baseband of the residual is extracted by a lowpass filter. The baseband signal is then decimated and waveform coded. The decimation rate is related to the cutoff frequency of the lowpass filter. The cutoff is usually selected to be in the range of 800 to 1000 Hz with a decimation rate of 4 or 5. Such a baseband signal includes the fundamental frequency and, hence, the pitch information. Another waveform-excited LPC is the *voice-excited vocoder* (VEV) which is very similar to RELP. The major difference between the RELP and VEV is that the VEV uses the baseband speech signal, rather than the residual signal, as the basis for its excitation waveform.

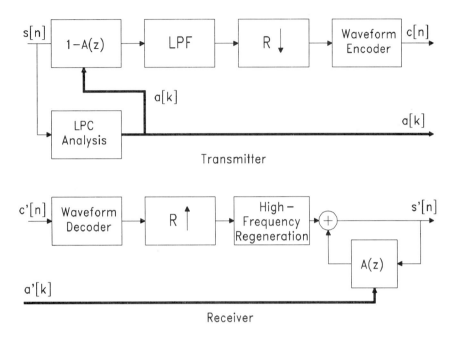

Figure 6.10. Block diagram of a residual-excited LPC (RELP) vocoder.

In the RELP synthesizer, the baseband residual is decoded and used to generate the excitation signal. Because the baseband residual is available at the receiver (because it is coded and transmitted), only the high-frequency information needs to be regenerated. Thus a critical component of RELP synthesis is the high-frequency regeneration algorithm that is used to produce the total excitation signal. Simple high-frequency generation methods include using nonlinearities such as square-law and rectification to spread energy into the upper bands while conserving the harmonic (pitch) structure. More complex methods include *spectral folding* and *spectral translation* in which the baseband spectrum is repeatedly copied onto higher frequency ranges. These methods are briefly discussed below.

Simple Nonlinearity Simple square-law or rectification of the baseband signal generates high frequencies from a baseband signal but does not provide an excitation with a flat spectrum. Rectification followed by a differencing operation provides a simple method to generate a relatively flat excitation signal for the LPC synthesizer.

Spectral Translation In spectral translation, the frequency-shifted versions of the baseband spectrum are used to generate the excitation signal. In practice, this can be achieved in the frequency domain by taking the DFT of the baseband signal, translating it properly, and taking the inverse DFT of the resulting full-band signal. Because the harmonic structure of the residual is not preserved by spectral translation, distortion tones often result.

Spectral Folding Spectral folding is very similar to spectral translation, but it reverses the spectrum of every second copy of the baseband spectrum. Spectral folding can be obtained by simply upsampling the baseband signal to the original sampling rate of the speech signal. As in spectral translation, spectral folding does not preserve the harmonic structure of the residual. The advantage of spectral folding is that it is very simple to implement.

EXERCISE 6.5.1. **Residual-excited LPC**

In this exercise, the role of the downsampling rate and the type of waveform coder used on the performance of RELP is studied. RELP analysis and synthesis is a four-part process. First, create the LPC coefficients and LPC residual. Second, generate a decimated residual using the RELP Analysis program. You may need to design a proper decimation filter using the filter design function. Third, code the decimated residual using a waveform coder. Finally, synthesize the output speech using the RELP Synthesizer program.

a. Using the high female (*h_female.sig*) and low male (*l_male.sig*) talkers, run the RELP analysis function with a decimation rate of 4. To do this, first generate the LPC coefficients and LPC residual (Signal Coding/Speech and Audio Coding/LPC/LPC Residual). Then, using the provided quarter-band FIR filter (*1to4.flr*), create a decimated residual using the RELP Analysis function (Signal Coding/Speech and Audio Coding/LPC/RELP Analysis). Then, without coding the excitation, run the RELP synthesis (Signal Coding/Speech and Audio Coding/LPC/RELP Synthesis) with both high-frequency regeneration algorithms. Listen to the synthetic speech and compute SNR (Signal Coding/Speech and Audio Coding/LPC/SNR S/N Ratio) for both cases.

b. Using an APCM with a step size adaptive quantizer (Signal Coding/Speech and Audio Coding/Waveform /APCM Step Size Adapt. Quant.), code the excitation signal using 2, 3, 4, 5, and 6 bits/sample and run the RELP synthesis again. What are the bit-rates of these systems? Comment on the quality of the synthetic speech. Compare RELP results to that of pitch-excited LPC and APC.

c. Repeat the above experiments with a decimation factor of 5. Note that you will have to design your own decimation filter.

6.6 Exercises

EXERCISE 6.6.1. Order of LPC and LTP Analyses

The order in which LPC and LTP analyses are performed on the speech signal may have some effect on the coder performance. In this exercise, the importance of this order is studied for the open-loop case.

a. Perform a 10th-order LPC analysis on one of the sentences and obtain the residual signal (Signal Coding/Speech and Audio Coding/LPC/LPC Residual). Perform an LTP analysis (Signal Coding/Speech and Audio Coding/LPC/LTP Analysis) with the standard parameters and save the residual signal in a separate file.

b. Change the order of the LPC and LTP analysis in part a and save the residual in another file. Compare the two residuals using the Statistics (Signal Utils/Signal Statistics) and Histogram (Signal Utils/Signal Histogram) functions, and by listening to them. Decide which one has more information and which one is more like white noise.

EXERCISE 6.6.2. LTP Analysis Parameters

The LTP analysis frame size and long-term memory size are the parameters of the open-loop LTP analysis. The effect of these parameters in the LTP analysis and the resulting long-term delay and gain are studied in this exercise.

a. Vary the long-term analysis frame size (Signal Coding/Speech and Audio Coding/LPC/LTP Analysis) within the allowed interval to see if the long-term delay has a pattern that matches the pitch period in voiced sections. You can do this by converting the LTP coefficient file (Edit/Convert Signal File) into an ASCII file and observing the delay values. Every other number in the file is the delay.

b. Write a small program that will read this ASCII file and save every other sample in a separate file. Convert this new file into an ASPI format and display it (Edit/Convert Signal File). Do you see the pitch pattern?

c. Change the long-term memory size and repeat the above sequence. What values of long-term memory produce the best pattern of pitch?

d. Find the frame size and long-term memory size that result in LTP residual with minimum variance.

EXERCISE 6.6.3. Other Quantization Techniques

All of the coders in this chapter allow the use of fixed, adaptive step size, and Jayant adaptive quantizers. Choose any of the previous exercises in this chapter, repeat them for different quantization techniques, and report your results.

Analysis-by-Synthesis LPC 7

7.1 Introduction

Analysis-by-synthesis speech coders, which include such important classes of coders as *code-excited linear predictive* (CELP) coders and *multipulse-excited linear predictive coders* (MPLPC), are among the newest and most effective of modern speech coders. All analysis-by-synthesis vocoders belong to the class of *waveform-excited* vocoders, and they use most of the available information from the speech signal to improve the quality and reduce the bit-rate. In particular, they make use of aural noise-masking, aural frequency resolution, aural phase insensitivity, syllabic energy variation, the long-term vocal tract properties, the short-term vocal tract properties, and pitch information in the coding process. In general, this makes analysis-by-synthesis coders among the best-quality, lowest bit-rate, and, as we shall see, most computationally demanding of speech coding techniques. Because analysis-by-synthesis coders make such extensive use of the classes of information available in the speech signal, we would expect these coders to also not be robust in the presence of bit errors, acoustic noise, multiple talkers, or nonspeech signals. To some extent, this is true. However, the analysis-by-synthesis model has proven remarkably flexible, and it has been shown to be more robust than might be expected.

Analysis-by-synthesis techniques form the basis for a flexible representation of the excitation signal in linear predictive vocoders. In this chapter, the principles of the analysis-by-synthesis procedure are discussed and some well-known LPC vocoders based on analysis-by-synthesis techniques are introduced. Although the residual excitation in LPC vocoders results in a considerable improvement of the speech quality over pitch-excited LPCs, the inherent problems in high-frequency regeneration techniques makes them unsatisfactory for high-quality speech ap-

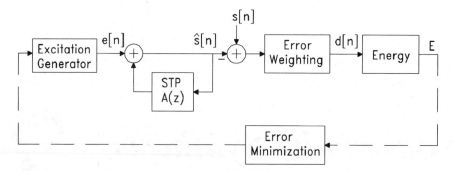

Figure 7.1. Block diagram of the analysis procedure used in analysis-by-synthesis linear predictive coders.

plications. As in RELP, analysis-by-synthesis-based excitations do not require voiced/unvoiced classification and pitch detection. On the other hand, analysis-by-synthesis provides a procedure to obtain a complete excitation signal without any need for high-frequency generation.

Analysis-by-synthesis coders can be thought of as either pitch-excited LPCs with a more effective, albeit higher bit-rate, excitation model; or as waveform excited LPCs that use a very efficient block-coding technique for encoding the residual signal. All analysis-by-synthesis coders represent the excitation signal using a small number of parameters, usually two to six. The excitation analysis procedure is performed by synthesizing speech using each possible set of parameters, and choosing the best by comparing it to the original speech using a perceptually based weighting function. In analysis-by-synthesis coders, each excitation parameter is normally the index in an ensemble of excitation functions. Therefore, the analysis-by-synthesis procedure is basically a way of exhaustively searching through an ensemble of alternatives for an optimum excitation sequence. Two important components of the analysis-by-synthesis procedure are the error-weighting function and the long-term predictor. As in noise-feedback coding, an error-weighting filter is used to help mask the coding noise, and the long-term predictor is used to exploit the pitch-periodic nature of voiced speech.

A block diagram of the analysis procedure for analysis-by-synthesis vocoders is shown in Figure 7.1. The parameters of the short-term predictor are obtained through the linear predictive analysis of speech as described in Chapter 5. In analysis-by-synthesis coders, the excitation generator is capable of generating K (usually 64-1024) different excitation sequences, $e_k(n)$. The analysis procedure generates all K separate possible speech signals, $\hat{s}_k(n)$, subtracts the original speech signal, and computes the weighted energy in the error signal. The analyzer must thus do K separate synthesis operations to choose the optimal excitation sequence; hence the name *analysis-by-synthesis*. In terms of the model described by Figure 7.1, the various analysis-by-synthesis LPCs are differentiated by the ensemble of sequences that are used to represent the excitation. The excitation sequences can be roughly divided into pulse excitation as in the multipulse-excited LPC (MPLPC), pitch excitation as in the *self-excited vocoder* (SEV), and code excitation as in code-excited LPC (CELP).

7.2 Excitation Model

Analysis-by-synthesis is a coding process where the excitation signal is determined on a block-by-block basis. It is assumed that the excitation signal for each block can, in general, be a combination of different excitation components as

$$e[n] = \sum_{k=1}^{M} \beta_k e_k[n] \tag{7.1}$$

where $e_k[n]$ is the kth excitation component. For the coders discussed in this chapter, the excitation components can be one of three types: a pulse, a codebook sequence, or the output of a long-term (pitch) predictor. The number of individual excitation components, M, is usually small. Typical examples are CELP, which has two components (a codebook sequence and a long-term predictor); SEV, which has one component (a long-term predictor); and MPLPC, which might have two to eight components (all pulses). Each excitation component is specified by an index, γ_k, and a corresponding gain, β_k. The optimum index for the kth component, γ_k, is determined by the analysis-by-synthesis procedure. For a given k and a given ensemble of K excitation sequences, $\mathcal{F}_k = \{f_\gamma[n], \gamma = 1, \ldots, K\}$, $e_k[n]$ is chosen to be the $f_{\gamma_k}[n]$ that minimizes the difference between the synthetic and original speech sequences in a weighted mean-squared sense. The optimum index, γ_k, and the associated gain value, β_k, are transmitted so that the excitation sequence can be reconstructed at the receiver.

Parameters associated with components of the excitation sequence are obtained in a suboptimum sequential method. This is accomplished by removing the effects of the previous excitation components from the original speech before obtaining the next component parameters. The motivation for this suboptimal approach is the prohibitive complexity of finding the optimal parameters for all excitation components. For example, finding the optimum pulse parameters in a multipulse-excited LPC results in a set of nonlinear equations in pulse locations and amplitudes. Although nonlinear equations may be solved by iterative procedures, they are usually very complex and not practical for real-time coders.

7.3 Error Weighting

An important component of the analysis-by-synthesis linear predictive coder is the error-weighting filter that is used to distribute the energy of the coding error signal in an appropriate manner. As in the noise-feedback coder (NFC) described in Section 6.4, error spectrum shaping in analysis-by-synthesis systems attempts to

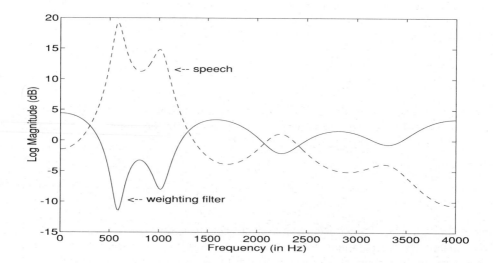

Figure 7.2. The speech spectrum for a voiced segment and the frequency response of the corresponding error-weighting filter with $\alpha = 0.8$.

mask the coding error signal (the masked signal) with the speech signal (the masking signal). The perceived loudness of the coding error signal is determined both by the overall power of the error signal and the spectral distribution of this signal with respect to the original speech. When the noise spectrum is flat, the perceived noise is in those regions of the spectrum where speech has low energy. Shaping the noise spectrum so that it is proportional to the signal spectrum reduces the perceived noise and therefore improves the overall speech quality.

In our implementation of analysis-by-synthesis LPC, the error-weighting filter by Atal and Schroeder is used [11]. This is an LPC-based block-adaptive weighting filter that is given by

$$W(z) = \frac{1 - A(z)}{1 - A(\alpha^{-1}z)} \quad (7.2)$$

where $0 \leq \alpha \leq 1$ is the weighting constant, and $A(z)$ is the short-term predictor. The filter, $W(z)$, emphasizes the error in speech spectral valleys and de-emphasizes it in formant regions. Figure 7.2 shows the spectrum of a voiced segment of speech and the magnitude response of the corresponding error-weighting filter, $|W(f)|$, for $\alpha = 0.8$.

An important advantage of the error-weighting filter, $W(z)$, as defined in Equation 7.2 is that its zeros cancel the LPC synthesis filter poles. The impulse response of the combined filter,

$$H_w(z) = H(z)W(z) = \frac{1}{1 - A(\alpha^{-1}z)}, \quad (7.3)$$

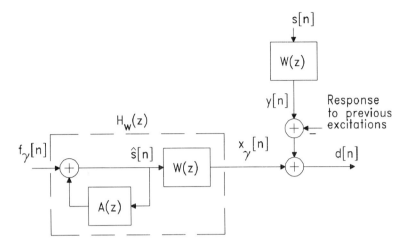

Figure 7.3. Block diagram of a source analysis model for a generic class of analysis-by-synthesis predictive coders. $s[n]$ is the input speech signal.

which is referred to as the weighted LPC synthesis filter, can be well approximated by a relatively short FIR filter,

$$h_w[n] \approx 0, \quad n > \ell \tag{7.4}$$

where ℓ is typically 20. $h_w[n]$ is referred to as the *weighted impulse response*. Such an approximation results in substantial computational savings in the analysis-by-synthesis search process.

7.4 Analysis-by-Synthesis Procedure

Assuming the multi-component excitation model of Equation 7.1, analysis-by-synthesis is used to obtain the index and gain of each excitation component. Figure 7.3 shows the block diagram of this analysis procedure for a generic class of analysis-by-synthesis predictive coders. Each component of the excitation sequence is obtained by minimizing the energy of $d[n]$, whose z-transform is given by

$$D(z) = Y(z) - \beta(\gamma) X_\gamma(z). \tag{7.5}$$

$Y(z) = S(z)W(z)$ is the z-transform of the weighted original speech, and $X_\gamma(z)$ is the z-transform of the weighted system response to the given excitation sequence, $f_\gamma[n]$. Figure 7.3 shows the block diagram to obtain $d[n]$. For any given γ, the corresponding gain, $\beta(\gamma)$, can be obtained by minimizing the weighted mean squared error given by

$$E_\gamma = \sum_{n=0}^{N-1} d^2[n] = \sum_{n=0}^{N-1} \left[y[n] - \beta(\gamma) \sum_{i=0}^{\ell-1} h_w[i] f_\gamma[n-i] \right]^2 \tag{7.6}$$

where $h_w[n]$ is the weighted impulse response. By equating the derivative of E with respect to β to zero, the expression for $\beta(\gamma)$ is given by

$$\beta(\gamma) = \frac{\sum_{n=0}^{N-1} y[n]x_\gamma[n]}{\sum_{n=0}^{N-1} x_\gamma^2[n]} \tag{7.7}$$

where

$$x_\gamma[n] = \sum_{i=0}^{\ell-1} h_w[i] f_\gamma[n-i] \tag{7.8}$$

is the weighted system response to the given excitation function, $f_\gamma[n]$. The associated mean squared error is given by

$$E(\gamma) = \left(\sum_{n=0}^{N-1} y[n]^2\right) - \frac{\left[\sum_{n=0}^{N-1} x_\gamma[n]y[n]\right]^2}{\sum_{n=0}^{N-1} x_\gamma[n]^2}. \tag{7.9}$$

The optimum index, γ_k, for the excitation component function, $e_k[n]$, is obtained by minimizing the mean squared error, E, over the allowable values of γ for the particular excitation ensemble being used.

Given any excitation ensemble, Equations 7.7-7.9 constitute a general method for selecting the component excitation sequence, $e_k[n] = \beta_k f_{\gamma_k}[n]$, that minimizes the perceptually weighted distortion. To obtain the parameters of the additional excitation components, the required procedure is to remove the effects of the previously determined component sequence, and perform analysis-by-synthesis on the remaining signal.

7.5 Long-Term Predictors

In linear predictive coders, the short-term redundancies of the speech signal (those due to the acoustic filtering effect of the vocal tract filter) are removed by using a short-term predictor. As in adaptive predictive coders, the long-term pitch-periodic redundancy of the voiced portions of the speech signal can also be exploited in analysis-by-synthesis coders. Figure 7.4 shows the LPC synthesizer with two cascaded synthesis filters. The short-term predictor (STP), $A(z)$, models the formant structure in the speech waveform, and the long-term predictor (LTP), $B(z)$, models the fine harmonic structure of speech. Sometimes, the LTP is also referred to as the pitch predictor. The general form of the long-term predictor is

$$B(z) = \sum_{i=-N_1}^{N_2} \beta_i z^{-\gamma-i} \tag{7.10}$$

where β_i's are the long-term predictor coefficients (gains), and γ is the long-term predictor delay. The number of coefficients is usually chosen to be from one to

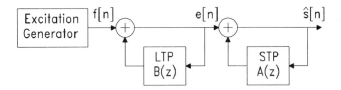

Figure 7.4. Block diagram of a generic analysis-by-synthesis LPC synthesizer with long-term predictor.

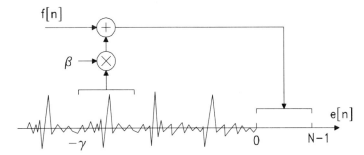

Figure 7.5. Search ensemble construction for the LTP. The optimum sequence is scaled by β and is used to update the search ensemble.

three. The delay, γ, is updated at the same rate as the predictor coefficients. In this text, we apply single-tap long-term predictors which are of the form $\beta z^{-\gamma}$.

The LTP can be considered to be an excitation source whose output is a component of the total excitation. The total excitation can be expressed as

$$e[n] = \beta e[n - \gamma] + f[n] \tag{7.11}$$

where $\beta e[n - \gamma]$ is the excitation component generated by the LTP, and $f[n]$ is the sum of all other excitation components. The long-term predictor parameters are obtained through an analysis-by-synthesis procedure in which γ is the index and β is the corresponding gain parameter. The search ensemble is a finite set of previous excitation sequences that can be represented by the set

$$\mathcal{F} = \{e[n - \gamma], \gamma = d_1, \ldots, d_2\} \tag{7.12}$$

where d_1 and d_2 determine the range of possible delay values that correspond to the expected range of the pitch period in speech, roughly from 5 to 20 msec. Figure 7.5 shows how the LTP search ensemble is constructed after each analysis frame. The ensemble is actually the memory of the LTP, and each ensemble function is a sequence of N samples beginning at sample $n = -\gamma$. Hence the LTP delay, γ, is the index of an ensemble whose sequences are formed by sliding a rectangular window across the memory of the LTP. The criterion for choosing the optimum delay is the minimum weighted error criterion mentioned earlier.

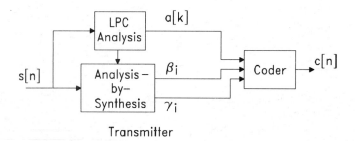

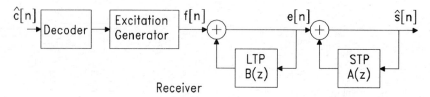

Figure 7.6. Transmitter and receiver of the MPLPC.

7.6 Multipulse-excited LPC (MPLPC)

In multipulse-excited LPC (MPLPC), the excitation signal is modeled as the sum of weighted, delayed impulses. Figure 7.6 shows a simple block diagram of an MPLPC transmitter and receiver. In the transmitter side, the speech signal is analyzed to generate the LPC coefficients, the LTP parameters, and the source parameters. The source parameters consist of the pulse locations and amplitudes. Thus the multipulse excitation can be represented as

$$f[n] = \sum_{k=1}^{M} \beta_k \delta[n - \gamma_k] \tag{7.13}$$

where γ_k's represent pulse locations and β_k's are the associated gains (or pulse amplitudes). For MPLPC, the total excitation sequence is given as

$$e[n] = \beta_0 e[n - \gamma_0] + f[n] = \beta_0 e[n - \gamma_0] + \sum_{k=1}^{M} \beta_k \delta[n - \gamma_k] \tag{7.14}$$

where β_0 and γ_0 are the LTP parameters and γ_k and β_k, where $k = 1, \ldots, M$, are M pulse locations and amplitudes. Based on the excitation model of Equation 7.1, the LTP and each pulse in the excitation model are considered to be a separate excitation component whose parameters (location and amplitude) are obtained using analysis-by-synthesis.

In analysis-by-synthesis of speech, the LTP parameters are determined first and the contribution of the LTP is subtracted from the original speech signal. Then,

parameters	name	range	typical values
predictor order	P	1–16	10
LPC window length	L	160-360	240
LPC frame size	I	80-240	160
error-weighting factor	α	0–1	0.8
LTP (excitation) frame size	N	20-60	40
number of pulses exc. frame	M	2-20	8

Table 7.1. The parameters of the MPLPC analysis and synthesis.

after the parameters of each pulse are determined, its contribution is subtracted from the original speech signal and the remaining signal is used to find the parameters of the next pulse. For a constant bit-rate coder, the number of pulses per frame is usually predetermined. It is also possible to adjust the number of pulses to meet a predetermined error criterion, which usually results in a variable bit-rate coder.

The parameters of the MPLPC consist of the parameters for the LPC analysis (window length, L; LPC frame size, I; predictor order, P); the error-weighting factor, α; the source analysis frame size, N; and the number of pulses per analysis frame, M. Table 7.1 gives the range and some typical values for these parameters.

Subjective evaluation results have shown that very good quality speech can be produced by the MPLPC in the bit-rate range of 9.6 to 16 kbps [7, 12]. The effectiveness of such a model at medium bit-rates lies in its ability to model both the periodic structure and the random nature of the LPC residual signal. Experiments also show that the LTP has a very significant effect on the performance of the MPLPC.

The parameters that are coded for transmission consist of the LPC coefficients, the LTP delay and gain, and the pulse location and amplitudes. Because of the number of parameters to be quantized, the performance of the MPLPC at bit-rates below 9.6 kbps decreases significantly. In order to reduce the number of parameters, locations of the pulses can be fixed in the analysis frame, which results in the regular pulse excitation model.

EXERCISE 7.6.1. MPLPC

To run analysis-by-synthesis LPC vocoders, you need to run the LPC Analysis function to generate the LPC coefficients; weighted impulse response, $h_w[n]$; and the weighted speech signal (Signal Coding/Speech and Audio Coding/LPC/Complete LPC Analysis). The basic multipulse excitation parameters are provided by the MPLPC function. Notice that the LPC frame size, I, must be an integer multiple of the excitation frame size, N.

a. Using the high female ($h\_female.sig$) and low male ($l\_male.sig$) talkers, create the basic LPC parameters (Signal Coding/Speech and Audio Coding/LPC/Complete LPC Analysis). You can use the standard parameters to do this, except for the pre-emphasis parameter which should be set to zero. Then run the MPLPC (Signal Coding/Speech and Audio Coding/LPC/MPLPC Multipulse LPC) with the standard parameters. Listen to the output speech and record its SNR. Reduce the number of pulses from 8 to 6, 4, 2, and 1. Compute the SNR for each case and listen to the results.

b. Reduce the excitation frame size, N, to 60 and repeat part a. How many pulses are enough for the $N = 60$ case to generate the same quality as $N = 40$ and $M = 8$?

c. Using the standard parameters, set $M = 0$. This means that the LTP is the only source of the excitation signal. This is the simplest version of what is known as the self-excited vocoder (SEV).

d. Run the MPLPC with $M = 1$ and $M = 8$ without any LTP by setting the long-term memory size to zero. Display the excitation signal and original speech signal. Can you observe any pattern for the location of the excitation pulses? Listen to your result. How good is MPLPC without an LTP?

7.7 Regular Pulse-excited LPC (RPLPC)

RPLPC is a special case of the multipulse excitation model in which the excitation pulses are equally spaced in time. In RPLPC, the first pulse can be located anywhere in the analysis frame, but the number of pulses in a frame and the spacings between them are entirely determined. Thus, after determining γ_1, other pulse locations are $\gamma_1 \pm kD$ where D is the space between the pulses in the excitation sequence. The coded parameters for the regular pulse excitation are the first pulse location, γ_1, and the pulse amplitudes, β_k, $k = 1, \ldots, \frac{N}{D}$, where N is the excitation analysis frame length.

The analysis procedure involves choosing the first pulse location and the associated gain values simultaneously to minimize the weighted error energy. The first pulse location is obtained using an analysis-by-synthesis procedure, and all pulse magnitudes are obtained by solving a set of N/D linear equations. High-quality speech is reported with a pulse spacing of $D = 4$ with an analysis frame size of 40 at around 9.6 kbps. Because of predetermined pulse spacings, RPLPC is computationally less complex than regular MPLPC.

EXERCISE 7.7.1. RPLPC

The amount of quality loss by positioning the excitation pulses at certain positions is usually small. In this exercise, the RPLPC is compared to the MPLPC.

a. Using the high female (*h_female.sig*) and low male (*l_male.sig*) talkers, create the basic LPC parameters (Signal Coding/Speech and Audio Coding/LPC/Complete LPC Analysis). You can use the standard parameters to

parameters	name	range	typical values
predictor order	P	1–16	10
LPC window size	L	160-360	240
LPC frame size	I	80-240	120
error-weighting factor	α	0–1	0.8
LTP (excitation) frame size	N	20-60	40
pulse separation	D	2-10	4

Table 7.2. The parameters of the RPLPC analysis and synthesis.

do this, except for the pre-emphasis parameter which should be set to zero. Then run the RPLPC (Signal Coding/Speech and Audio Coding/LPC/Regular Multipulse LPC) with the standard parameters. Listen to the output speech and record its SNR. Reduce the number of pulses from 8 to 5, 4, 2, and 1. Compute the SNR for each case and listen to the results.

b. Compare the results with the MPLPC with the same number of pulses per frame. How does RPLPC compare to MPLPC in SNR? How does RPLPC compare to MPLPC in quality?

7.8 Code-excited LPC (CELP)

The MPLPC has proven to be a very effective speech coder at medium bit-rates (9.6 kbps and above). A major portion of the bits in MPLPC are used to code the excitation parameters, the pulse location, and the amplitudes. The number of bits for the excitation coding must be significantly reduced to achieve lower bit-rates. In RPLPC, the number of pulse locations is reduced to one. In order to reduce the bit-rate below 9.6 kbps, the number of excitation parameters must be reduced even further. The code excitation model is very effective in modeling the excitation with a very low number of parameters. The code-excited linear predictive coder (CELP) [7, 12] is another analysis-by-synthesis based speech coder for good quality at bit-rates below 9.6 kbps. As in the MPLPC, short-term and long-term predictors are used to model the spectral envelope and pitch-periodic structure of the speech signal.

In the code excitation model, a relatively large codebook of random or deterministic code sequences (codewords) is used for excitation modeling. In CELP, the total excitation can be expressed as

$$e[n] = \beta_0 e[n - \gamma_0] + \beta_1 v_{\gamma_1}[n] \qquad (7.15)$$

where $v_\gamma[n]$, $\gamma = 1, \ldots, K$, $n = 0, \ldots, N-1$ is the N-sample sequence in the codebook with index γ, where K is the codebook size. N is the size of the analysis frame, which is usually chosen to be around 5 msec. Increasing the analysis frame

138 Analysis-by-Synthesis LPC

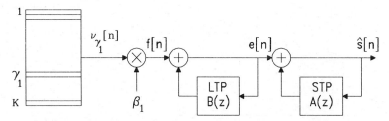

Figure 7.7. Block diagram of a CELP synthesizer.

parameters	name	range	typical values
predictor order	P	1–16	10
LPC window size	L	160-360	240
LPC frame size	I	80-240	120
error-weighting factor	α	0–0.99	0.8
codeword (exc. frame) size	N	20-60	40

Table 7.3. The parameters of the CELP analysis and synthesis.

size usually results in a reduction in the coder performance. A typical codebook size is 1024, which requires $\log_2 1024 = 10$ bits for representation.

In the transmitter, an analysis-by-synthesis procedure is used to obtain the optimum codeword. The codeword that results in the minimum weighted error energy is selected as the optimum sequence. The index of the optimum codeword and the corresponding scaling factor are coded and used to generate the excitation sequence in the synthesizer. Figure 7.7 is the block diagram of a CELP synthesizer.

The main disadvantage of code excitation is the high computational cost of the searching process. Most of the computational load in CELP comes from the exhaustive search of the codebook by filtering each of the candidate sequences through the synthesis filters. About 500 million multiply-add operations per second are required for a codebook with 1024 entries of length 40. Various procedures have been proposed for efficient searching of the codebook. Some of them, such as fast search using singular-value decomposition (SVD), result in optimum sequences with up to an order of magnitude savings in the computation. Many other procedures provide computational savings by applying a preprocessing method. In preprocessing, a subset of the codebook is selected and the selected subset is exhaustively searched. In our implementation of CELP, a long random sequence instead of a codebook is used. The long sequence can be searched very efficiently using a recursive search algorithm. In this algorithm, the contribution of the sample at the end of the codeword is subtracted, and the contribution of the new sample at the beginning of the codeword is added to the filtered response. Table 7.3 summarizes the parameters of the CELP coder and their typical values.

EXERCISE 7.8.1. **CELP**

a. Using the high female (*h_female.sig*) and low male (*l_male.sig*) talkers, create the basic LPC parameters (Signal Coding/Speech and Audio Coding/LPC/Complete LPC Analysis). You can use the standard parameters to do this, except for the pre-emphasis parameter which should be set to zero. Then run the CELP (Signal Coding/Speech and Audio Coding/LPC/Code Excited LPC (CELP)) with the standard parameters. Listen to the output speech and record its SNR. Compare this result to the MPLPC and RPLPC at the same frame rates.

b. Change the excitation frame rate to 20 and 60. Compute the SNR for each case and listen to the results. How does the quality of the CELP change with the size of the excitation frame size?

c. Try codebook sizes of 64, 128, 512, and 1024. Compute the SNR and listen to the results. How is the quality of the CELP related to codebook size?

d. Vary the weighting factor between 0.0 and 0.9 and observe the difference. Weighting factors larger than 0.95 will result in a substantial amount of degradation. This is because the approximation of the weighted impulse response by a short FIR filter, $h_w[n]$, will not be valid anymore.

7.9 Exercises

EXERCISE 7.9.1. **Predictor Order and LPC Frame Size**

A common feature to all analysis-by-synthesis LPC vocoders is the LPC analysis. In this exercise, the effect of the LPC parameters such as predictor order, analysis frame size, and window length is studied. For this experiment, you can use any of the analysis-by-synthesis coders (MPLPC, RPLPC, CELP, or SEV).

a. Using the standard LPC parameters (Signal Coding/Speech and Audio Coding/LPC/Complete LPC Analysis) with $M = 5$ and $N = 20$, change the predictor order from 1 to 16 in a few steps. Then apply the coder of your choice. How does the quality vary with the predictor order? How about SNR (Signal Coding/Speech and Audio Coding/LPC/SNR S/N Ratio)? Repeat this exercise with $N = 40$.

b. With a fixed predictor order of 10, window length of 240, and $N = 40$, change the LPC frame size from 40 to 80, 120, 160, 200, and 240. Describe the behavior of the vocoder. How does the SNR change?

EXERCISE 7.9.2. **Multiple Speaker and Noisy Speech**

An interesting experiment is to test the performance of analysis-by-synthesis coders with multiple talkers and a noisy speech signal.

a. Using the Scale (Signal Utils/Signal Scale) and Add functions (Signal Utils/Signal Addition), generate a multiple-talker signal. You may want to make one of them weaker by scaling it. A mixture of a male talker and a female talker may be a good choice. You may also need to scale both signals to prevent clipping.

b. Run the Complete LPC Analysis (Signal Coding/Speech and Audio Coding/LPC/Complete LPC Analysis) and then run the MPLPC (Signal Coding/Speech and Audio Coding/LPC/Multipulse LPC (MPLPC)) and the CELP (Signal Coding/Speech and Audio Coding/LPC/Code Excited LPC (CELP)) with the standard parameters. Listen to the output speech and compute their SNRs (Signal Coding/Speech and Audio Coding/LPC/SNR S/N Ratio). How do these coders perform with multiple talkers? Would you prefer the waveform coder performance to the performance of analysis-by-synthesis coders?

c. Generate a noise signal (Signal Utils/Random Signal) and then add the noise (Signal Utils/Signal Addition) to a female and a male talker sentence. At different SNR levels such as 0, 10, and 20 dB, run the Complete LPC Analysis program (Signal Coding/Speech and Audio Coding/LPC/Complete LPC Analysis) and then run MPLPC (Signal Coding/Speech and Audio Coding/LPC/Multipulse LPC (MPLPC)) and the CELP (Signal Coding/Speech and Audio Coding/LPC/Code Excited LPC (CELP)) with standard parameters. How do the coders perform with noisy signals?

EXERCISE 7.9.3. Music

Music signals have very different characteristics from speech signals. In this exercise, performance of the analysis-by-synthesis vocoders is studied for music signals.

a. Using one of the available music signals, run the Complete LPC Analysis program (Signal Coding/Speech and Audio Coding/LPC/Complete LPC Analysis) and then run MPLPC (Signal Coding/Speech and Audio Coding/LPC/Multipulse LPC (MPLPC)) and the CELP (Signal Coding/Speech and Audio Coding/LPC/Code Excited LPC (CELP)) with standard parameters. How do the coders perform? Compare the performance of the MPLPC and the CELP to the performance of the pitch-excited LPC on music.

b. Mix a music file and a speech file with the music as the background noise. Try the CELP and the MPLPC on this signal.

Subband Coding

8.1 Introduction

In all the coders that were considered in previous chapters, the speech signal is processed in the time domain and the signal is treated as a full-band signal. Such coders are referred to as *time-domain coders*. This chapter is concerned with a different class of coders called *frequency-domain coders*. In *frequency-domain coders*, the speech signal is first transformed into an alternate *time-frequency* representation, and it is this alternate representation that is used as the basis for the coding process.

The basic subband coding and transform coding process is illustrated in Figure 8.1. The speech signal is first transformed by the *analysis* section into its time-frequency representation by either a filter bank or a block transform such as a discrete cosine transform. All analysis techniques used for speech coding are *maximally decimated*. This means that the number of samples in the time-frequency representation is exactly equal to the number of samples in the time-domain representation. The samples of the time-frequency representation are then coded using the *channel coder* and decoded using the *channel decoder*. The output of the decoder then goes to the *synthesis* section, which reconstructs the coded speech using a synthesis filter bank or inverse block transform. If the channel coder and channel decoder are removed, then the remaining system is called the *analysis-synthesis* system. An analysis-synthesis system is said to be *exactly reconstructing* if, in the absence of channel coders and decoders, it is capable of exactly reconstructing the input signal at the output.

As was discussed in the introduction, a major advantage of frequency-domain coders is that they can directly incorporate knowledge about the human auditory

142 Subband Coding

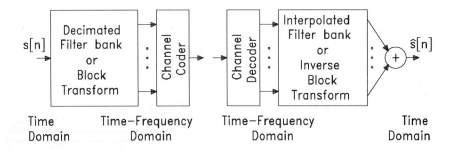

Figure 8.1. Basic frequency-domain coder.

system in ways that are not possible using time-domain coders. From a speech coding perspective, the most important property of the ear is the *noise-masking* effect in which the ear intrinsically masks noise components that are close in frequency to signal components. Because the coding of any signal always generates noise in the reconstructed signal, the ear's noise-masking effect can sometimes be used to make coding noise inaudible.

It is possible to take some advantage of the noise-masking effect in time-domain coders. For example, in such time-domain coders as APC and analysis-by-synthesis LPC coders, noise shaping is used to take advantage of aural noise masking using error-weighting filters. However, frequency-domain coders have an advantage because the perceived quantization noise can be minimized by shaping its spectrum directly in the frequency domain. The frequency shaping of the quantization noise is achieved by using a different number of bits to encode each frequency component. Thus frequency bands with low energy can be coded with only a few bits, and bands with very low energy may not be assigned any bits at all.

Two major types of frequency-domain speech coders are subband coders (SBC) and adaptive transform coders (ATC). In SBCs, the speech is divided into four or more frequency bands and each of these *subband signals* is coded separately. In subband coders, the quantization noise generated in a particular band is largely confined to the band in which it was generated. Bands with high signal energies usually have high corresponding quantization levels as well. Since this noise will generally be masked by the large signal in the same band, it will be less audible than if it were allowed to spread into bands with relatively less signal energy. Separate adaptive quantizers can be applied in different bands, and adaptive bit allocation schemes can be used to distribute the available bits among the different bands to minimize the perceived coding noise.

In adaptive transform coders, the speech is divided into blocks, each speech block is transformed, and the transform coefficients are coded. Each transform coefficient can be coded independently of the other coefficients according to its importance. In point of fact, subband coders and transform coders are members of the same class of coders. The primary difference is that transform coders usually have more channels and poorer equivalent channel filters than subband coders. Because of the time-varying nature of the speech spectrum, adaptive bit allocation schemes are usually more effective than fixed bit allocation schemes.

In this book, basic concepts of frequency-domain coders are presented in the context of subband coders.

8.2 Subband Coding

Figure 8.2 shows the block diagram of a basic subband coder. In subband coding, the speech is first split into frequency bands using a bank of bandpass filters. The individual bandpass signals are then decimated and encoded for transmission. At the receiver, the channel signals are decoded, interpolated, and added together to form the received signal.

The important parameters in subband coders are the number of frequency bands and the frequency coverage of the system, and the way subband signals are coded. Uniform-band structures, where all the bands have equal widths, and octave-band structures, where the bandwidths are half as great as the higher adjacent band and twice as great as the lower adjacent band, are the most common frequency splits used in subband coders. Subband signals are usually coded using APCM (see Chapter 3) and low-frequency bands are usually coded with more bits to achieve a better perceived speech quality.

The subband coder derives its quality advantage by limiting the quantization noise from the encoding/decoding operation largely to the band in which it is generated, thereby taking advantage of known properties of aural perception. This results in a noise shaping similar to that of the APC and analysis-by-synthesis LPC coders.

To control the spread of the coding distortion, the quality of the bandpass filters in the analysis-synthesis system is an important factor in subband coder performance. In addition, issues such as system delay and the amount of aliasing distortion involved in the analysis-synthesis filter banks are relevant in subband coder design.

8.2.1 Two-Band Analysis-Synthesis Systems

If not properly designed, analysis-synthesis filter bank systems result in significant distortion in the reconstructed speech. The distortion consists of spectral magnitude, spectral phase, and aliasing distortions that are to be minimized or completely eliminated by designing proper analysis and reconstruction filter banks.

Many existing subband coders employ tree-structured analysis-synthesis filter bank structures. The basic component of tree-structured subband coders is the two-band analysis-reconstruction system shown in Figure 8.3. In this system, the analysis is performed by the two frequency selective filters, $H_0(e^{j\omega})$ and $H_1(e^{j\omega})$, which are nominally half-band lowpass and half-band highpass filters respectively. In order to preserve the system sampling rate, both channels are maximally decimated at a rate of two-to-one, resulting in the two subsampled signals, $Y_0(e^{j\omega})$ and

144   Subband Coding

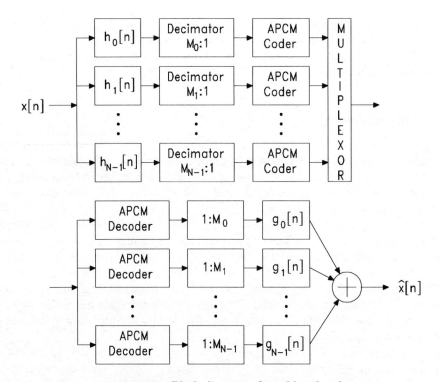

Figure 8.2. Block diagram of a subband coder.

$Y_1(e^{j\omega})$, given by

$$Y_0(e^{j\omega}) = (1/2)[H_0(e^{j\omega/2})X(e^{j\omega/2}) + H_0(-e^{j\omega/2})X(-e^{j\omega/2})] \qquad (8.1)$$

$$Y_1(e^{j\omega}) = (1/2)[H_1(e^{j\omega/2})X(e^{j\omega/2}) + H_1(-e^{j\omega/2})X(-e^{j\omega/2})]. \qquad (8.2)$$

In the reconstruction section, the bands are recombined, giving

$$\hat{X}(e^{j\omega}) = (1/2)[H_0(e^{j\omega})G_0(e^{j\omega}) + H_1(e^{j\omega})G_1(e^{j\omega})]X(e^{j\omega}) \\ +(1/2)[H_0(-e^{j\omega})G_0(e^{j\omega}) + H_1(-e^{j\omega})G_1(e^{j\omega})]X(-e^{j\omega}). \qquad (8.3)$$

The frequency response of the two-band linear system component is contained in the first term of Equation 8.3, while the second term contains the aliasing. In the quadrature mirror filter (QMF) solution, the aliasing is removed by defining the reconstruction filters as

$$G_0(e^{j\omega}) = H_1(-e^{j\omega}) \qquad (8.4)$$

$$G_1(e^{j\omega}) = -H_0(-e^{j\omega}) \qquad (8.5)$$

This assignment forces the aliasing to zero, and results in a total system frequency response, $C(e^{j\omega})$, of

$$C(e^{j\omega}) = (1/2)H_0(e^{j\omega})H_1(-e^{j\omega}) - (1/2)H_1(e^{j\omega})H_0(-e^{j\omega}) \qquad (8.6)$$

Sec. 8.2 Subband Coding   145

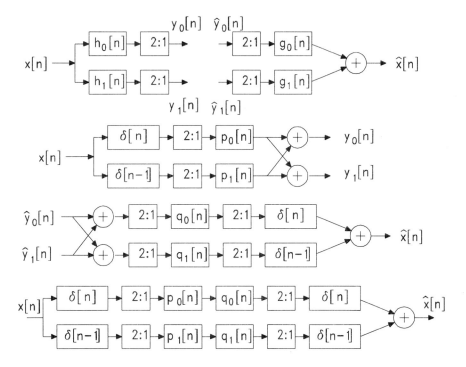

Figure 8.3. Two-band subband analysis-reconstruction systems.

In the QMF solution, the highpass filter, $H_1(e^{j\omega})$, and lowpass filter, $H_0(e^{j\omega})$, are chosen to be frequency-shifted versions of each other by π, so that

$$H_1(e^{j\omega}) = H_0(-e^{j\omega}). \tag{8.7}$$

For this class of analysis-reconstruction systems, exact reconstruction requires that

$$H_0^2(e^{j\omega}) - H_1^2(e^{j\omega}) = 2. \tag{8.8}$$

A number of authors using various methods have designed FIR filters that approximate this condition. Other authors have reported filter constraints that satisfy the perfect reconstruction conditions of the system [13]. Conjugate quadrature filters (CQF) are defined to satisfy the following conditions:

$$H_1(e^{j\omega}) = -H_0(-e^{-j\omega}) \tag{8.9}$$
$$G_0(e^{j\omega}) = H_0(e^{-j\omega}) \tag{8.10}$$
$$G_1(e^{j\omega}) = H_1(e^{-j\omega}). \tag{8.11}$$

Like the QMF solution, the CQF solution removes the aliasing distortion, and the total system frequency response, $C(e^{j\omega})$, can be expressed as

$$C(e^{j\omega}) = (1/2)F(e^{j\omega}) + (1/2)F(-e^{j\omega}) \tag{8.12}$$

where

$$F(e^{j\omega}) = H_0(e^{j\omega})H_0(e^{-j\omega}) \qquad (8.13)$$

is called the product filter. The condition for exact reconstruction in this case is that

$$F(e^{j\omega}) + F(-e^{j\omega}) = 2. \qquad (8.14)$$

Such product filters are relatively easy to design, and $H_0(z)$ can be obtained using spectral decomposition techniques.

Recently, a number of methods have been reported for the design of M-band analysis-reconstruction filter banks [13], and nonuniform-band filter banks with perfect reconstruction. Based on these systems, analysis-synthesis systems with a better match to the critical band model of the ear can be obtained. However, in this text, all subband coders are based on tree-structured systems mentioned earlier in this section. More specifically, QMFs, CQFs, and some filters with aliasing and spectral distortions are provided for experiments.

EXERCISE 8.2.1. Two-Band Analysis System

In this exercise you will examine the output of a two-band analysis system. The two-band analysis system is the fundamental component of all tree-structured analysis-synthesis systems, and thus of octave-band subband coders.

a. For the high female (*h_female.sig*) and low male (*l_male.sig*) talkers, perform a two-band analysis (Signal Coding/Speech and Audio Coding/Subband/2-Band Sub. Analysis) for 8-tap (F8.FB2), 16-tap (F16.FB2), and 32-tap (F32.FB2) quadrature mirror filters. Listen to the outputs (File/Playback Signal) of the individual channels and describe what you hear. Can you hear the input speech signal in the individual channels? Why?

b. Do a long-term spectrum analysis (Signal Utils/Spectrum Analysis) of the high-band and low-band signals and plot the result. How do you explain the slope of the two spectra?

c. Find the value of the optimal predictors for the high and low bands. Why is the predictor value for the high band negative?

d. Based on the results of c, would fixed differential coders be effective in a two-band subband coder?

EXERCISE 8.2.2. Two-Band Synthesis System

The most basic function of an analysis-synthesis system is to reconstruct a good approximation of the input signal at the output. In this exercise, you will demonstrate the ability of a two-band analysis-synthesis system to reconstruct the input signal. To do this exercise, you will need the results from the previous exercise.

a. Using the high and low output signals from the previous exercise (you should have six of each, one high female (*h_female.sig*) and one low male (*l_male.sig*) talker for 8-tap, 16-tap, and 32-tap systems), reconstruct the six associated full-band signals (Signal Coding/Speech and Audio Coding/Subband/2-Band Sub. Synthesis) and listen to them. Can you hear any distortions?

b. Graphically compare the input (original) and output (reconstructed) signals and determine the delay between the signals. Can you see any distortion?

c. Using your measured delay to control the offset of the SNR function, compute an SNR (Signal Coding/Speech and Audio Coding/Subband/SNR S/N Ratio) for the six signals.

d. Based on your results, are there any differences between short (lower-quality filters) and long (higher-quality filters) analysis-synthesis systems?

EXERCISE 8.2.3. **Two-Band Subband Coder**

In this exercise, you will generate two two-band subband coders at 32 and 16 kbps, and compare their performance to a full-band APCM coder.

a. Using an APCM coder (Signal Coding/Speech and Audio Coding/Waveform/APCM Step Size Adapt. Quant.) with standard parameters, code the high female (*h_female.sig*) and low male (*l_male.sig*) talkers at 16 and 32 kbps.

b. Using the two-band subband coding function (Signal Coding/Speech and Audio Coding/Subband/2-bnd Subband Coder), code the high female (*h_female.sig*) and low male (*l_male.sig*) talkers at 16 (C4.CC2) and 32 (C16.CC2) kbps using 8-tap (F8.FB2) and 32-tap (F32.FB2) analysis systems.

c. Do a careful listening test in which you compare and rank all 32-kbps systems with respect to one another (full band, 8-tap subband, and 32-tap subband). Does a two-band subband coder system offer any advantage at 32 kbps? How does the performance of the different two-band subband coders compare?

d. Do a careful listening test in which you compare and rank all 16-kbps systems with respect to one another (full band, 8-tap subband, and 32-tap subband). Does a two-band subband coder system offer any advantage at 16 kbps? How does the performance of the different two-band subband coders compare?

8.2.2 Tree-Structured Analysis-Synthesis Systems

Tree-structured analysis-synthesis systems concatenate two-band analysis systems to make multiband analysis systems. In a similar way, two-band synthesis systems are concatenated to form corresponding multiband synthesis systems.

In this text, the multiband examples are all six-band octave-band analysis-synthesis systems. The basic elements of such systems are a two-band *split* and a two-band *merge* function, as shown in Figure 8.4. The *split* function consists of two analysis filters, $h_0(n)$ and $h_1(n)$, each of which is followed by a 2:1 decimator.

148   Subband Coding

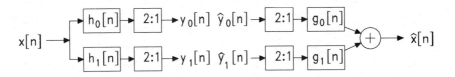

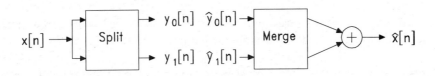

Figure 8.4. Split and Merge functions for tree-structured subband coders.

Thus, for every two samples in the time domain, $x(n)$ and $x(n+1)$, the *split* function produces two time-frequency samples, $y_0(\frac{n}{2})$ and $y_1(\frac{n}{2})$. The *merge* function consists of two 2:1 upsamplers followed by two synthesis filters, $g_0(n)$ and $g_1(n)$. The outputs of the two synthesis filters are then added to produce the reconstructed signal, $\hat{x}(n)$.

The way in which these systems are combined to make six-band analysis-synthesis systems is shown in Figure 8.5. *Split A* divides the full band into two half-bands. If the speech is sampled at 8000 samples per second, then the two are 0-2000 Hz and 2000-4000 Hz. The low-frequency band is then split again, giving two bands at 0-1000 Hz and 1000-2000 Hz. This process is repeated five times to give the six-band frequency partition shown in Figure 8.5. The *delay* elements in the nonpartitioned high-frequency bands are needed to compensate for the extra delay added to the partitioned channels by the subsequent analysis-synthesis systems.

The synthesis systems perform the inverse of the analysis operation. The *Merge E* function begins by combining the two lowest bands, 0-125 Hz and 125-250 Hz, into the reconstructed 0-250 Hz band. This process is then repeated until the full-band signal, $\hat{x}(n)$, is reconstructed. Once again the *delay* elements are needed to compensate for the reconstruction delay of the bands that have been split.

EXERCISE 8.2.4. Six-Band Analysis-Synthesis System

A six-band octave-band system will provide an appreciable aural noise-masking effect. Such a subband coder system is implemented by the Subband Coder function and three different filter sets: the 32 − 32 − 32 − 32 − 32 tap set (F32.FB6), the 8−8−8−8−8 tap set (F8.FB6), and the 32−16−16−16−8 tap set (FX.FB6). In this exercise you will investigate how these systems work in the absence of quantization

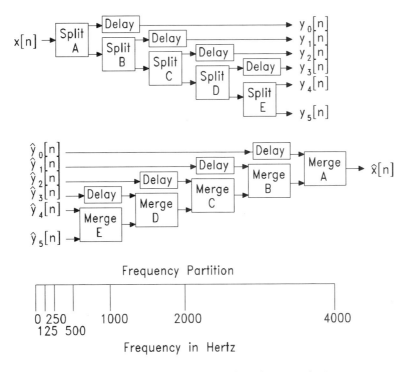

Figure 8.5. Six-band tree-structured analysis-synthesis system.

errors (noise).

a. For the high female (*h_female.sig*) and low male (*l_male.sig*) talkers, perform a subband analysis (Signal Coding/Speech and Audio Coding/Subband/6-bnd Sub. Analysis) and synthesis (Signal Coding/Speech and Audio Coding/Subband/6-bnd Sub. Synthesis) for the three available filter banks and listen to them. Can you hear any distortions?

b. Graphically compare the input (original) and output (reconstructed) signals and determine the delay between the signals. A good approach is to start with a wide view of all four signals and then use the Measure command (Display/Measure) to make a rough estimate of the delay. Then make the view narrow and fine tune your estimate. Can you see any distortion?

c. Using your measured delay, compute an SNR (Signal Coding/Speech and Audio Coding/Subband/SNR S/N Ratio) for the six signals.

d. Based on your results, what are the important differences between short (lower-quality filters) and long (higher-quality filters) octave-band analysis-synthesis systems?

8.2.3 Subband Signal Coding

The coding of subband signals is another important part of a subband coder. PCM, APCM, and ADPCM coders with fixed or adaptive bit-allocation schemes can all be used in subband signal coding. Adaptive bit-allocation schemes usually result in better coder performance than fixed bit-allocation with a higher computational complexity. In a fixed bit allocation scheme, the number of bits assigned to each band is fixed at all times. On the other hand, in a subband coder with adaptive bit-allocation, the available bits are distributed among channels to maximize the quality of the coded speech.

In subband coders used in this text, APCMs are based on work by Jayant [6]. The adaptive quantizers in these systems are controlled by the dynamic step size, $\Delta(n)$, given by

$$\Delta(n) = \Delta(n-1) F[c(n-1)] \qquad (8.15)$$

where $c(n)$ is the nth code word and $F[\ ]$ is a preset control function. The control parameters for the subband coders are provided in two files. The first file includes the analysis and synthesis filter coefficients and the second file includes the quantizer parameters. Refer to the help files on the subband coders in DSPLAB.

EXERCISE 8.2.5. **The Subband Coding System**

In this exercise, you will compare the effectiveness of the different available octave-band analysis-synthesis systems for reducing perceived coding distortion. There are three available analysis-synthesis systems and three available bit-rates: 24 kbps, 20 kbps, and 16 kbps.

a. For the high female (*h_female.sig*) and low male (*l_male.sig*) talkers, use the octave-band subband coder (Signal Coding/Speech and Audio Coding/Subband/6-bnd Subband Coder) to code at 24 kbps (C24.CC6), 20 kbps (C20.CC6), and 16 kbps (C16.CC6) for the 32 − 16 − 16 − 16 − 8 (FX.FB6) tap analysis-synthesis system. Also generate an ADPCM-coded (Signal Coding/Speech and Audio Coding/Waveform/ADPCM Step Size Adpt. Quant.) version of the same sentences at 32 kbps and 16 kbps. You should use a first-order fixed predictor with a predictor value between 0.8 and 0.9.

b. Do a careful listening test and rank the systems from part a. What does this tell you about the ability of the octave-band subband coding system to mask coding noise?

c. Now repeat part a for the 32 − 32 − 32 − 32 − 32 tap (F32.FB6) and the 8 − 8 − 8 − 8 − 8 tap (F8.FB6) analysis-synthesis systems.

d. Do a careful listening test and rank the subband coding systems at the same rate as to their ability to mask coding noise. What does this tell you about the importance of filter quality for masking coding noise?

8.3 Exercises

Product coding is a technique that allows the use of non-integer numbers of bits for individual channels in a subband coder. If, in a six-band octave-band subband coder, each channel were to use an integer number of bits, say b_1 to b_6, then the total number of bits for 32 samples (16 samples from channel 6, 8 from channel 5, 4 from channel 4, 2 from channel 3, and 1 each from channels 2 and 1) would be $16b_6 + 8b_5 + 4b_4 + 2b_3 + b_2 + b_1$. If the quantizers are parameterized in terms of the number of levels they can represent rather than the number of bits, and if a channel uses l_i levels, it would require

$$b_i = log_2(l_i) \tag{8.16}$$

bits. Using product codes, it is possible to take advantage of quantizers in which the number of levels is not a power of 2, and hence the number of bits is not an integer.

In a product code, a number of samples are coded together using a fixed number of bits. In a very simple case, assume that two quantized channel outputs with l_1 and l_2 levels respectively were to be coded together. Then the required number of bits would be

$$bits = \lceil log_2(l_1) + log_2(l_2) \rceil = \lceil log_2(l_1 l_2) \rceil \leq \lceil log_2(l_1) \rceil + \lceil log_2(l_2) \rceil \tag{8.17}$$

where $\lceil \ \rceil$ is the ceiling operator. For the 32-sample, six-band example above, the number of bits would be

$$bits = \lceil log_2(l_1) + log_2(l_2) + log_2(l_3) + log_2(l_4) + log_2(l_5) + log_2(l_6) \rceil \tag{8.18}$$

Thus, using product codes, it is possible to assign bits more flexibly than using an integer numbers of bits.

EXERCISE 8.3.1. Product Codes

In this exercise, you will realize several different subband coders using integer and non-integer numbers of bits per channel, and compare the results.

a. Using the high female (h_female.sig) and low male (l_male.sig) talkers, realize a six-band 16-kbps subband coder in two ways: first, using 4 levels (2 bits) (C4.CC6) in all channels; second, using the level pattern of $l_6 = 3$, $l_5 = 4$, $l_5 = 6$, $l_5 = 8$, $l_5 = 9$, $l_5 = 8$ (CX.CC6) for the FX.FB6 filter file.

b. Compare the two subband coders to one another and to several 16-kbps APCM and ADPCM systems. Which system is best?

c. Design your own 16-kbps system by choosing alternate sets of levels that can be realized at 16 kbps. Can you improve the performance over the standard system?

EXERCISE 8.3.2. Delay Compensation

As was discussed in this chapter, any subbands that are not split in a tree-structure must be delayed in order to achieve correct reconstruction. In this exercise, you will experiment with the effects of delay compensation.

a. Using the high female (*h_female.sig*) and low male (*l_male.sig*) talkers, realize a six-band analysis-synthesis system using the 32-tap filter set (F32.FB6).

b. The first parameter in the filter control file (F32.FB6) is the delay compensation for the lowest band. For a 32-tap filter set, this value is 31. Create three new filter control files (U1.FB6, U2.FB6, and U3.FB6) that are exactly the same as F32.FB6 except that the lowest band compensations are set to 1, 15, and 30 respectively. Now repeat part a using your new control files.

c. Listen to the results from part b. What does the distortion sound like? What does this tell you about the sensitivity of octave-band subband coders to correct delay compensation?

EXERCISE 8.3.3. Aliasing Distortion

Aliasing distortion in speech and audio subband coding, if not cancelled properly, usually results in very noticeable degradation of coded speech. In this exercise you will design some analysis and synthesis filters and create aliasing distortion.

a. Using the FIR design function, design a half-band filter, $H_0(z)$. Using Equations 8.4, 8.5, and 8.7, obtain other system filters. Implement the two-band systems. Listen to the output without quantizing the subband signals. Do you hear any distortion? Notice that this system has no aliasing or phase distortion.

b. Using the Spectrum Analysis function of DSPLAB, obtain the frequency response of the transfer function of the two-band system. Measure the peak spectral magnitude distortion.

c. Instead of using Equation 8.5, use $G_1(e^{j\omega}) = H_0(-e^{j\omega})$ to obtain $G_1(z)$. Run the two-band system and listen to the output. How different does the output sound compared with your result in part b? Notice that in this system, not only do the two aliasing components not cancel each other, but they are added together to emphasize the aliasing distortion.

d. You can also use unrelated analysis and synthesis filters to observe the effect of aliasing. Use 8-tap analysis and 16-tap synthesis QMF filters and listen to the output. How much aliasing is audible?

Projects 9

A project is considered to be a major undertaking, and not a simple exercise. A course on speech coding, for example, would normally involve only one or two projects. Some of the projects can be performed within the DSPLAB environment and some require extra programming. Whenever extra programming is needed, the results generated by the user can be saved in an ASCII format and the Convert function of the DSPLAB can be used to change the parameters to an ASPI file format. An example of an ASCII signal file can be obtained by converting a DSPLAB file (which has an ASPI file format) to an ASCII file using the Convert function.

Following is a list of projects that can be used throughout the course.

Project 9.1: *Dithering*

The concept of dithering is used to decorrelate the quantization noise from the speech signal [6]. In dithering, a dither signal, usually a high-frequency random signal, is added to the signal before quantization.

a. To generate the dither signal you can use the random sequence generator in DSPLAB, which is located in the signal utility menu of DSPLAB. After generating the random sequence, you can Filter and Scale to generate the dithering signal.

b. Add the dithering signal to the original signal and use PCM with a low number of bits (1 or 2). You then may or may not subtract the dithering signal from the output. Compare the results of PCM with dithering to regular PCM. Try different dithering signals. You can use the functions in the signal utility menu of DSPLAB to generate different dithering signals. For example, use high-frequency filter with different cutoffs to generate different high-frequency dithering signals. You can also change the level of the dithering signal by using the Scale function and study its effects.

c. Listen to the results of PCM with dithering. You may also listen to the coding noise of PCM with dithering and compare it to the coding noise generated by the regular PCM. You may also study the effect of subtracting the dithering signal from the coded signal.

Project 9.2: *Laplacian Quantizer*

a. Using Equation 3.11, find the optimum Laplacian compressor function with at least 30 points. Use the derived data in the General Companding PCM function of DSPLAB and test it on at least three different speech signals with 1, 2, 3, 4, and 5 bits. The General Companding PCM is under the Waveform menu. Compare your subjective and objective results to μ-law and A-law results.

b. Using the quantizer data given in Table 3.3, write a program, in the language of your choice, that would implement the optimum Laplacian nonuniform quantizer with 1, 2, 3, 4, and 5 bits. Use your program to quantize the speech signals you worked with in part a and compare your results to the results of part a for the same bit-rates.

Project 9.3: *PCM-ASB*

Using the programming language of your choice, implement a feedback step size adaptive PCM (PCM-ASB). To implement such a system, you need to program three major components. First, write a program that estimates the speech energy using the window of your choice. Second, implement the step size adaptation rule based on the speech energy. Some rules for the step size adaptation are presented in Chapter 3. Finally, implement a fixed uniform quantizer. You can compare your results to that of the programs provided in DSPLAB.

Project 9.4: *Coding of Noisy Speech*

In this project, you will perform a complete analysis of the performance of PCM, log-PCM, APCM, and ADPCM with noisy speech.

a. To perform this experiment, you need to add different amounts of noise to speech to achieve SNRs such as 0, 5, 10, 15, and 20 dB. You can use the functions under the Signal Utils menu to perform this task. Use the random number generator to obtain the noise signal. You can scale this signal and add it to the speech signal and measure the SNR.

b. Evaluate the adaptive PCM performances for different SNR values. For example, sensitivity of the adaptive quantizers to noisy speech may be studied. An adaptive quantizer with instantaneous and syllabic adaptation may respond in different ways to the noisy speech.

c. The effectiveness of differential coding for noisy speech can be evaluated. Find the optimum first-order fixed predictor coefficients for the noisy speech at different SNRs (at least 4 SNRs of 0, 5, 10, and 20 dB) and compare them to the coefficients of the clean speech and pure noise signals. Make a graph of the predictor coefficient value with respect to SNR (from $-\infty$ for pure noise to $+\infty$ for pure speech.) You have six data points to make your graph. Is it possible to approximate the predictor coefficient value for an SNR of 15 dB from this graph? Compute the predictor coefficient for the 15 dB SNR and compare that to your prediction.

d. What can you say for the predictor value that you computed for a pure noise signal in part c? What would you think the predictor value would be for a constant signal? Create a constant signal using the signal utilities of DSPLAB with a magnitude of 5000 and length 10000. Find the fixed predictor for this signal using the LPC menu functions. Did you guess the predictor value correctly? Explain the results.

Project 9.5: *DPCM with Adaptive Quantization*

Write a program that implements ADPCM with a Jayant quantizer (Figure 4.3). The program involves three major parts. The first part is the implementation of the predictor and the quantization loop. The second part is the step size adaptation for the Jayant quantizer. The third part is to obtain the optimum fixed predictor, which can be performed using the fixed predictor function under the LPC menu of DSPLAB. Compare the results of your program with the results of the ADPCM function in DSPLAB.

Project 9.6: *Predictor Coefficient Quantization*

The performance of APC in real applications can be evaluated more realistically with quantized predictor coefficients. In this project, the quantization of predictor parameters is considered. Using the Convert Signal File function, the file that contains the predictor coefficients can be converted into an ASCII file. Any ASCII file can be converted into ASPI format using the same function. LPC coefficients are saved as FLOAT_32.

a. Pick a sentence and perform LPC analysis on it. Convert the LPC coefficient file and the PARCOR file into ASCII files.

b. Write a program that will quantize the predictor parameters. Your choices are numerous. You can quantize LPC coefficients directly, which may not be a very good choice, or any other equivalent representation of these coefficients as listed in Chapter 5. If you generate an ASCII file with the same format as that of the original LPC ASCII file, you can convert it back to an ASPI format for the APC program to use.

c. Compare the performance of APC with fixed and adaptive quantizers, with and without the predictor coefficient quantization. Try to quantize LPC coefficients, PARCORs, LARs, and so forth, and compare them in terms of the performance of APC.

d. Compute the bit-rate and SNR in each case and compare the results to other coders. Listen to your results and try to detect the distortion created by predictor coefficient quantization.

Project 9.7: *Linear Predictive Coding*

A good set of projects for pitch-excited LPC vocoders is to write programs that compute basic LPC functions. The Convert function in the laboratory software can be used to interface the new software to the laboratory programs. Some good LPC related projects are listed below. Notice that each of the following items can be considered a major project.

- Write a new LPC analysis routine using the autocorrelation method. The input to your program should be an input speech signal, and the output should be a file that contains frames of data. Each frame should contain a gain followed by P LPC feedback coefficients, where P is the number of LPC coefficients in the analysis.
- Write a new LPC analysis routine using the covariance method. The input to your program should be an input speech signal, and the output should be a file that contains frames of data. Each frame should contain a gain followed by P LPC feedback coefficients, where P is the number of LPC coefficients in the analysis.
- Write a pitch detector program. The input should be a speech signal, and the output should be a set of integer pitch-period values, one for each 10-msec (80-sample) frame. Unvoiced speech would have a zero pitch-period value.
- Write a new LPC synthesizer program. The program should take an LPC coefficient file (see above) and a pitch file (see above) as input and give synthetic speech as output.
- Write a new LPC analyzer-synthesizer pair based on PARCOR coefficients rather than the feedback coefficients.
- Write a time-scale modification system using an LPC analyzer and synthesizer.

Project 9.8: *Predictor Parameter Quantization*

In all analysis-by-synthesis coders provided in DSPLAB, unquantized predictor coefficients are used to obtain the synthetic speech signal. In this project, LPC parameters are to be quantized and used in the provided functions. To do this task, perform the necessary LPC analysis functions and generate the LPC parameters.

Using the Convert function, and writing your own quantization program, quantize the LPC parameters. The quantized LPC parameters can then be used to generate the LPC synthetic speech. As mentioned in Chapter 5, LPC parameters can be represented in different formats such as PARCORs, LARs, and LSPs. The LPC analysis function in DSPLAB generates the LPC coefficients as predictor coefficients and PARCORs. In this project you need to quantize the LPC coefficients using at least three different representation of the LPC parameters. Suggested representations include the predictor coefficients, PARCORs, LARs, and cepstrum coefficients (see Chapter 5).

Project 9.9: *Subband Coding*

A good set of projects for a subband coder is to write programs that compute basic subband coder functions. As usual, the convert functions in the laboratory software can be used to interface the new software to the laboratory system.

Subband coders are very modular. In an octave-band subband coding system, for example, there are only two basic components: a two-band analysis-synthesis system, and a channel coder. Thus, by writing only two programs, any number of subband coders can be simulated.

Some good subband coder projects are as follows:

- There are many published articles on analysis-synthesis systems based on filter banks. Many of these are on two-band and multiband filter banks for maximally decimated, exactly reconstructing analysis-synthesis systems. Many of the articles publish filter coefficients. Based on these, it is possible to create a number of analysis-synthesis systems for new subband coders. Some possible examples are:

    – Two-band quadrature mirror filter banks

    – Two-band conjugate quadrature (Smith-Barnwell) filter banks

    – Two-band linear phase filter banks

    – Two-band minimum delay filter banks

    – Multiband filter banks

    – Combinations of filter banks in tree-structures

- As with the analysis-synthesis systems, many different approaches can be used for channel coding. Many of these coders were discussed in Chapters 3 and 4. From the perspective of this text, a channel coder is a program that takes a signal in and generates a quantized signal out. Some possible channel coders include:

    – PCM

- Companded (log) PCM
- APCM
- ADPCM
- Block APCM

- By combining new analysis-synthesis systems with new channel coders, create new subband coders.

Menu Items

This appendix provides a listing of all the menu items as well as an alphabetical list of all the signal-related functions. The first three tables (Tables A.1, A.2, and A.3) list the items available under the File, Edit, and Display menus. These three menus are always directly available from any of the other menus. The Waveform, APC, LPC, and Subband menus (Tables A.4, A.5, A.6, and A.7) can be reached through the Speech and Audio Coding menu. The Signal Utils and Window FIR menus (Tables A.8 and A.9) are the remaining menus that are available from the main screen. There are no submenus for the Signal Utils and Window FIR menus.

A.1 Menu Items by Menu Name

This section lists all the menu items on a menu-by-menu basis.

Menu Items

File
Configure DSPLAB
Change Directory
Delete File(s)
Memory Usage
Save Screen
Restore Screen
Restore Screen Settings
Print Signal Plot
Playback Signal
Record Signal
Execute Ashell Command
DOS Shell
Help Print
Quit DSPLAB

Table A.1. Menu items under the File menu.

Edit
Edit file(s)
Signal file info
Modify signal file
Convert signal file
Undo
Cut Signal
Copy Signal
Paste Signal
Save Segment to File
Play Segment
Playback Signal
Record Signal

Table A.2. Menu items under the Edit menu.

Display
Locked Next Frame
Locked Previous Frame
Go to Frame n (Locked)
Lock Windows
Next Frame
Previous Frame
Open Multiple Windows
Change Multiple Windows
Open Window
Change Window Parameters
Close Window
Close Current Window
Close All Windows
Measure
Zoom In
Zoom Out
Setup Windows
Select Window
Full Screen
Tiled Screen
Redraw Screen
Fast Display Menu

Table A.3. Menu items under the Display menu.

Waveform
PCM Uniform
Log PCM Mu-Law
Log PCM A-Law
General Companding PCM
APCM Step Size Adapt. Quant.
APCM Gain Adapt. Quant.
APCM Jayant Quantizer
DPCM Uniform Quantizer
ADPCM Jayant Quantizer
ADPCM Step Size Adapt. Quant.
LDM Delta Modulation
ADM Adaptive Delta Modulation
CVSD Continuously Variable Slope DM
Optimum Fixed Predictor
SNR S/N Ratio
Segment S/N Ratio (SEGSNR)
Signal Statistics
Signal Histogram

Table A.4. Menu items under the Waveform menu.

APC
Optimum Adaptive Predictor
APC Uniform Quantizer
APC Jayant Quantizer
APC Step Size Adapt. Quant.
D*PCM Uniform Quantizer
D*PCM Jayant Quantizer
D*PCM Step Size Adapt. Quant.
APC-PP Uniform Quantizer
APC-PP Jayant Quantizer
APC-PP Step Size Adapt. Quant.
APC-NF Uniform Quantizer
APC-NF Jayant Quantizer
APC-NF Step Size Adapt. Quant.
APC-PPNF Uniform Quantizer
APC-PPNF Jayant Quantizer
APC-PPNF Step Size Adapt. Quant.
SNR S/N Ratio
Segment S/N Ratio (SEGSNR)

Table A.5. Menu items under the APC menu.

LPC
Simple LPC Analysis
Complete LPC Analysis
Excitation to Speech
Pitch Detection
LPC Vocoder (complete)
LPC Synthesis (pitch exc.)
LPC Residual
LTP Analysis
RELP Analysis
RELP Synthesis
MPLPC Multipulse LPC
Regular Multipulse LPC
CELP Code Excited LPC
SEV Self Excited Vocoder
SNR S/N Ratio
Segment S/N Ratio (SEGSNR)

Table A.6. Menu items under the LPC menu.

Subband
2-bnd Subband Coder
2-bnd Sub. Analysis
2-bnd Sub. Synthesis
6-bnd Subband Coder
6-bnd Sub. Analysis
6-bnd Sub. Synthesis
SNR S/N Ratio
Segment S/N Ratio

Table A.7. Menu items under Subband menu.

Signal Utils
Signal Addition
Signal Center Clipping
Signal Center Clip (2 sig.)
Signal Clipping
Signal Clipping (2 sig.)
Signal Constant
Signal Decimation
Signal Delay
Signal Energy
Signal Filter
Signal Interpolation
Signal Histogram
Signal ASCII Multiply
Signal Signal Multiply
Signal Ramp
Signal Scale
Signal Statistics
Signal Subtraction
Signal Zero Crossing
Spectrum Analysis
Compare Two Spectra
Waveform Compare
Random Signal
SNR S/N Ratio
Segment S/N Ratio (SEGSNR)

Table A.8. Menu items under the Signal Utils menu.

Window FIR
Lowpass
Bandpass
Plot Options

Table A.9. Menu items under the Window FIR menu.

A.2 Alphabetical List of Menu Items

This section lists the menu items in alphabetical order and shows where the menu items can be found.

Signal Menu Item	Menu
ADM Adaptive Delta Modulation	Waveform
(Signal) Addition	Signal Utils
ADPCM Jayant Quantizer	Waveform
ADPCM Step Size Adapt. Quant.	Waveform
APC Jayant Quantizer	APC
APC Step Size Adapt. Quant.	APC
APC Uniform Quantizer	APC
APCM Gain Adapt. Quant.	Waveform
APCM Jayant Quantizer	Waveform
APCM Step Size Adapt. Quant.	Waveform
APC-NF Jayant Quantizer	APC
APC-NF Step Size Adapt. Quant.	APC
APC-NF Uniform Quantizer	APC
APC-PP Jayant Quantizer	APC
APC-PP Step Size Adapt. Quant.	APC
APC-PP Uniform Quantizer	APC
APC-PPNF Jayant Quantizer	APC
APC-PPNF Step Size Adapt. Quant.	APC
APC-PPNF Uniform Quantizer	APC
(Signal) ASCII Multiply	Signal Utils
Bandpass	Window FIR
CELP Code Excited LPC	LPC

Table A.10. Signal-related menu items and their corresponding menu names (continued in Tables A.11 and A.12).

Signal Menu Item	Menu
(Signal) Center Clipping	Signal Utils
(Signal) Center Clip (2 sig.)	Signal Utils
(Signal) Clipping	Signal Utils
(Signal) Clipping (2 sig.)	Signal Utils
Compare Two Spectra	Signal Utils
Complete LPC Analysis	LPC
(Signal) Constant	Signal Utils
CVSD Continuously Variable Slope DM	Waveform
(Signal) Decimation	Signal Utils
(Signal) Delay	Signal Utils
DPCM Uniform Quantizer	Waveform
D*PCM Uniform Quantizer	APC
D*PCM Jayant Quantizer	APC
D*PCM Step Size Adapt. Quant.	APC
(Signal) Energy	Signal Utils
Excitation to Speech	LPC
(Signal) Filter	Signal Utils
General Companding PCM	Waveform
(Signal) Histogram	Signal Utils
(Signal) Interpolation	Signal Utils
LDM Delta Modulation	Waveform
Log PCM A-Law	Waveform
Log PCM Mu-Law	Waveform
Lowpass	Window FIR
LPC Synthesis (pitch exc.)	LPC
LPC Vocoder (complete)	LPC
LTP Analysis	LPC
MPLPC Multi Pulse LPC	LPC
Optimum Adaptive Predictor	APC
Optimum Fixed Predictor	Waveform
PCM Uniform	Waveform
Pitch Detection	LPC
Pitch Exc. LPC Synthesis	LPC
Plot Options	Window FIR
(Signal) Ramp	Signal Utils
Random Signal	Signal Utils

Table A.11. Signal-related menu items and their correponding menu names (continued from Table A.10).

Signal Menu Item	Menu
RELP Analysis	LPC
RELP Synthesis	LPC
(Signal) Scale	Signal Utils
Segment S/N Ratio (SEGSNR)	APC
Segment S/N Ratio (SEGSNR)	LPC
Segment S/N Ratio (SEGSNR)	Signal Utils
Segment S/N Ratio (SEGSNR)	Subband
Segment S/N Ratio (SEGSNR)	Waveform
SEV Self Excited Vocoder	LPC
(Signal) Signal Multiply	Signal Utils
(Signal) Statistics	Signal Utils
(Signal) Subtraction	Signal Utils
6-bnd Sub. Analysis	Subband
6-bnd Sub. Synthesis	Subband
6-bnd Subband Coder	Subband
SNR S/N Ratio	APC
SNR S/N Ratio	LPC
SNR S/N Ratio	Signal Utils
SNR S/N Ratio	Subband
SNR S/N Ratio	Waveform
Spectrum Analysis	Signal Utils
2-bnd Sub. Analysis	Subband
2-bnd Sub. Synthesis	Subband
2-bnd Subband Coder	Subband
Waveform Compare	Signal Utils
Window FIR	Window FIR
(Signal) Zero Crossing	Signal Utils

Table A.12. Signal-related menu items and their correponding menu names (continued from Table A.11).

Extending DSPLAB B

B.1 Adding a Custom D/A Driver

DSPLAB includes support for four sound boards/DSP cards. If you do not have one of these cards, you may still be able to playback signals. First, you need to create a PC executable program that can read and play a file using the D/A on your sound board. This program may have been supplied with your sound board, or you may have to write this program yourself using your sound board's documentation. This program should accept the name of the file to play and the sampling rate used to playback the file.

Once you have this program, you will need to edit the file D_TO_A.BAT in the BIN directory. Assuming that your sound board playback program is called **playback.exe** and can read a binary 16-bit signed integer data file directly (all the signal data files use a 16-bit signed integer format), the batch file on your monitor will look like the batch file shown in Figure B.1. As described in the next section, the DSPLAB signal files have a 128-byte header that preceeds the actual 16-bit signal data. You can either ignore this header (which will sound like a 'pop' in the beginning of your data), or you may write another program that will 'strip off' the 128-byte header before you playback the file.

Finally, you will need to run the Configure DSPLAB option under the file menu and select the CUSTOM_D_to_A option for the sound board.

B.2 ASPI Signal File Format

The file header associated with the ASPI signal files contains basic information about the contents and source of the file. The file header will always consist of a

```
rem  This batch file allows you to use your own D/A driver
rem  for playback of the signals.  This batch file will be
rem  called as follows:
rem
rem          D_TO_A.BAT filenameToPlayback  sampleRateInHertz
rem
rem  where filenameToPlayback is the name of the signal file to
rem  playback and sampleRateInHertz is the sample rate in Hertz
rem  to be used when playing back the signal.
rem

playback.exe %1 %2
```

Figure B.1. Example batch file for interfacing to your own D/A board.

block of 128 bytes. The file body contains all data. The header is structured as shown in Table B.1.

Valid data types for signal files are shown in Table B.2. The most common data type within DSPLAB is the 16-bit signed integer format.

B.3 Example C Program

An example C program, example.c, for reading and writing integer signal files is included with this book. This program allows you work with the signal files provided and to create your own signal files that are compatible with the plotters provided with DSPLAB. As a reminder, you can also create ASCII data files and convert them to signal files by using *Convert signal file* under the Edit Menu.

```
/******************************************************************

DSPLAB Example Program

This program is an example of reading an integer Signal file,
processing the data, and writing an output integer Signal file.
Integer Signal files are used because they produce smaller output
files than floating point Signal files (only two bytes per sample
rather than four) and are compatible with all the coding and signal
utility programs provided with DSPLAB.  Internally to this program
the integer data files are converted to floating point numbers for
convenience and then converted back to integers before being written
to the output signal file.
```

Sec. B.3   Example C Program   171

Field	# Bytes	Format	Value	Purpose
1	8	ASCII	ASPIFILE	Label to identify an ASPI file
2	4	32-int		Number of samples (may be zero)
3	2	16-int		Revision number (currently 2)
4	2	16-int	0	File type. Zero for signal file.
5	2	16-int	0 thru 8	Data format (see following list)
6	4	32-fp		Sample rate in Hertz
7	2	16-int	0 or 1	Normalization flag (0 for normalized, 1 for not norm.)
8	4	32-fp		Full-scale data range
9	46		Unused	
10	20	ASCII		Default output device
11	32	ASCII		Tag - a 32 character message
12	2	16-int		Negative of field 3

Table B.1. ASPI signal file header format.

```
To run this program, type:

example input_file output_file

where input_file is the input signal file and output_file is the
output signal file.
**********************************************************************/

#include <stdio.h>
#include <math.h>
#include <malloc.h>

#define BUFFER 256 /* Read 256 samples at a time */

/* Define the Signal header data structure */
typedef struct HDR HDR;
struct HDR
{
    char     aspifile[8];
    unsigned long nsample;
    short    revision;
    short    file_type;
    int      file_data_type;
    float    sample_rate;
```

	Data Format
0	16-bit signed integer
1	32-bit IEEE floating point
2	64-bit IEEE floating point
3	32-bit signed integer
4	8-bit unsigned integer
5	8-bit signed integer
6	16-bit unsigned integer
7	32-bit unsigned integer
8	C30 floating point

Table B.2. Valid data types for signal files.

```
   short    normalization;
   float    file_data_range;
   char     unused[46];
   char     device[20];
   char     descriptor[32];
   short    signature;
};
/***************************************************************
Clip the data to the desired range.
***************************************************************/
void limitData(float *in, float max, float min, int numSamples)
{
   int i;

   for (i=0;i<numSamples;i++)
   {
      if (in[i] > max)     in[i] = max;
      else if (in[i] < min) in[i] = min;
   }
}
/***************************************************************
Convert from floating point to integer data.
***************************************************************/
void floatToInt(float *in, int *out, int numSamples)
{
   int i;

   /* Convert to an integer */
   for (i=0;i<numSamples;i++)
   {
```

```c
      out[i] = (int) in[i];
   }
}
/***************************************************************
Convert from integer to floating point data.
***************************************************************/
void intToFloat(int *in, float *out, int numSamples)
{
   int i;

   /* Convert to an integer */
   for (i=0;i<numSamples;i++)
   {
      out[i] = (float) in[i];
   }
}
/***************************************************************/
main (int argc, char **argv)
{
   HDR inhead;
   int numRead, i, *inbuf, *outbuf;
   float *floatIn;
   FILE *in, *out;

   /* Open the input file */
   if (!(in=fopen(argv[1], "rb")))
   {
      printf("Error: Could not open input file %s\n", argv[1]);
      exit(1);
   }

   /* Open the output file */
   if (!(out=fopen(argv[2], "wb")))
   {
      printf("Error: Could not open output file %s\n", argv[2]);
      exit(1);
   }
   /* Allocate integer input arrays */
   inbuf = (int *)malloc(BUFFER*sizeof(int));
   outbuf = (int *)malloc(BUFFER*sizeof(int));

   /* Allocate floating point array */
   floatIn = (float *)malloc(BUFFER*sizeof(float));
```

```c
   if (!floatIn || !inbuf || !outbuf)
   {
      printf("Error: Could not allocate memory.\n");
      exit(1);
   }
   /* Read an input integer signal file */

   /* Read the 128-byte signal header */
   fread(&inhead, sizeof(char), 128, in);

   /* Write out the same header to the output */
   fwrite(&inhead,sizeof(char),128, out);

   /* Continue while there is data, numRead is the
      number of samples read */
   while (numRead = fread(inbuf,sizeof(int),BUFFER, in))
   {
      /* Convert to floating point data */
      intToFloat(inbuf,floatIn, numRead);

      /* Process the data, in this example simply negate the data */
      for (i=0;i<numRead;i++)
      {
         floatIn[i] *= -1.0;
      }
      /* Limit the data to fit into the valid output
         integer data range */
      limitData(floatIn, 32767.0, -32768.0, numRead);

      /* Convert to integer data for smaller output file */
      floatToInt(floatIn, outbuf, numRead);

      /* Write the output integer signal file */
      fwrite(outbuf,sizeof(int),numRead,out);
   }
}
```

Glossary of Abbreviations C

ADM	adaptive delta modulator
ADPCM	adaptive differential PCM
APC	adaptive-predictive coding
APC-APB	adaptive-predictive coding with feedback adaptive predictor
APC-APF	adaptive-predictive coding with feed-forward adaptive predictor
APC-NF	adaptive-predictive coding with noise feedback
APC-PP	pitch-predictive adaptive-predictive coding
APC-PPNF	pitch-predictive adaptive-predictive coding with noise feedback
APCM	adaptive pulse code modulation
CELP	code-excited linear predictive coder
CVSD	continuously varying slope delta modulator
D*PCM	open-loop DPCM
DM	delta modulator
DPCM	differential DPCM
DPCM-AQ	DPCM with adaptive quantization
DPCM-AQB	DPCM with feedback adaptive quantization
DPCM-AQF	DPCM with feed-forward adaptive quantization
FB	feedback
FF	feed-forward
LDM	linear delta modulator
log-PCM	logarithmic companding DPCM
LPC	linear predictive coding
LTP	long-term predictor

MPLPC	multipulse linear predictive coder
NFC	noise-feedback coding
PCM	pulse code modulation
PCM-AB	feedback adaptive DPCM
PCM-AF	feed-forward adaptive DPCM
PCM-AGB	feedback gain adaptive DPCM
PCM-AGF	feed-forward gain adaptive DPCM
PCM-ASB	feedback step size adaptive DPCM
PCM-ASF	feed-forward step size adaptive DPCM
RELP	residual-excited linear predictive coder
RPLPC	regular pulse-excited linear predictive coder
SBC	subband coder
SEGSNR	segmental signal-to-noise ratio
SEV	self-excited vocoder
SNR	signal-to-noise ratio
STP	short-term predictor
VEV	voice-excited linear predictive coder

Bibliography

[1] K. Chapman and G. NG, *Sound Blaster 16 User Reference Manual*. Creative Technology Ltd., 67 Ayer Raja Crescent #03-18, Singapore 0513, 2nd ed., June 1993.

[2] Atlanta Signal Processors, Inc., 1375 Peachtree Street NE, Suite 690, Atlanta, GA 30309-3115, *Elf Hardware Instruction Manual*, elfa-001-03/95 ed.

[3] J. L. Flanagan, *Speech Analysis Synthesis and Perception*. Springer-Verlag, 1972.

[4] N. Chomsky, *The Sound Pattern of English*. Harper & Row, 1968.

[5] L. R. Rabiner and R. W. Schafer, *Digital Processing of Speech Signals*. Prentice Hall, 1978.

[6] N. S. Jayant and P. Noll, *Digital Coding of Waveforms*. Prentice-Hall Signal Processing Series, Prentice-Hall, 1984.

[7] D. O'Shaughnessy, *Speech Communication: Human and Machine*. Addison-Wesley, 1987.

[8] J. M. Pickett, *The Sounds of Speech Communication*. University Park Press, 1980.

[9] J. D. Markel and A. H. Gray, *Linear Prediction of Speech*. Springer-Verlag, 1976.

[10] T. P. Barnwell, "Recursive windowing for generating autocorrelation analysis for lpc analysis," *IEEE Transactions on Acoustics, Speech, and Signal Processing*, vol. ASSP-29, October 1981.

[11] B. S. Atal and M. R. Schroeder, "Predictive coding of speech and subjective error criteria," *IEEE Transactions on Acoustics, Speech, and Signal Processing*, vol. ASSP-27, pp. 247–254, June 1979.

[12] J. R. Deller, J. G. Proakis, and J. H. L. Hansen, *Discrete-Time Processing of Speech Signals*. MacMillan, 1993.

[13] P. P. Vaidyanathan, *Multirate systems and filter banks*. Prentice Hall, 1993.

Index

Δ, 43, 44, 46, 76
Φ, 93
α, 72, 73, 75, 135, 137, 138
β, 114, 115, 122, 132, 133
γ, 114, 115, 131–133, 137
λ, 94–96
μ-law, 13, 49–51, 53, 65, 154
θ, 119–122

A-law, 11, 13, 49–51
A/D, 3, 9, 28, 31, 42
Acoustic, 88, 98, 101
    excitation, 4
    filter, 5
    noise, 12, 107, 127
    radiation, 67
    tube, 4
    vocal tract filter, 67, 85, 132
    wave, 2, 4, 5
Adaptation
    feed-forward, 54
    feedback, 61
    gain, 59
    Jayant, 63
    step size, 56
Adaptive predictor, 68, 72, 107, 108, 110
Adaptive quantizer, 41, 42, 53, 68, 72, 110, 111, 116, 120, 124, 125, 150, 155
Adaptive-Predictive Coding, 108
ADM, 11, 12
ADP.PAR, 19
ADPCM, 11, 13, 68, 72, 74, 75, 111, 122, 150, 154, 155, 158
Algorithm, 1, 2, 10, 17, 37, 72, 86, 92, 101, 123, 124, 138
Aliasing, 9, 11, 143–146
All-pole, 88, 90, 103
All-voiced, 105
Amplitude, 29, 30, 37, 44–47, 68, 87, 129, 134–137
Analog, 11, 13, 38, 75
Analysis, 1, 5, 6, 9, 10, 17, 32, 68, 71, 72, 128, 131, 133, 135–137, 139, 141, 143, 146–150, 154–157
    block size, 115
    excitation, 86
    frame, 91, 92, 95, 96
    frames, 86
    long-term, 104
    LPC, 85, 86, 89, 90, 94–96, 99–101, 103, 104, 108–110, 125

LTP, 115, 116, 125
  recursive, 110
  RELP, 124
  window, 92, 98
Analysis-by-synthesis, 85, 115, 127–140, 142, 143, 156
Analysis-reconstruction, 143, 145, 146
Analysis-synthesis, 141, 143, 146–150, 157, 158
Analyzer-synthesizer, 156
APC, 7, 11, 12, 68, 107–111, 115, 116, 119, 120, 122, 124, 142, 143, 155, 156
APC-APB, 108
APC-APF, 108, 110
APC-NF, 120, 122
APC-PP, 108, 114–116, 119, 120, 122
APCM, 42, 72, 74, 75, 124, 143, 147, 150, 154, 158
Articulators, 5
ASCII, 20, 28–30, 105, 125, 153, 155, 156
Ashell, 18, 25–29
ASPI, 19, 20, 25, 28–31, 125, 153, 155, 156
ATC, 7, 11, 12, 142
Audio, 3, 13, 38
Auditory, 42
Aural, 41, 47, 67, 127, 142, 143, 148
  model, 7
  noise-masking, 7, 11
  perception, 4, 7, 9, 11, 107
Autocorrelation, 91–96, 101, 102, 115, 156
  method, 91, 93
AUTOEXEC.BAT, 19, 20
Autoregressive, 89

Background, 140
Band-limited, 41
Bandpass, 143
Bands, 7, 11, 123, 142–144, 146, 148

Bandwidth, 1, 3, 7, 38, 86, 88, 119, 143
Bank, 1, 5, 7, 141, 143, 146, 149, 157
Baseband, 122–124
BIN directory, 19, 20, 169
Bit allocation, 150
Bit errors, 3, 10, 42, 109, 127
Block-coding, 128
BUFFERS
  in config.sys, 20

Canon BJ and LBP, 27
CCITT, 13
Cellular, 13
CELP, 12–14, 127, 129, 137–140
Cepstrum, 103
CGA, 19
Channel
  coder, 141
  decoder, 141
Children, 98
Classification, 122, 128
Clipping, 46, 101, 121, 140
Code excitation, 138
Codebook, 129, 137–139
Code excitation, 137
Codeword, 137, 138
Cognitive model, 3
Compand, 1, 11, 13, 42, 47, 49, 67
Compress, 1, 38, 42, 47, 49, 154
CONFIG.SYS, 18, 20
Conjugate, 145
Consonant, 97
Correlation, 67, 75, 86, 89, 103, 108, 115
Correlator, 87
Covariance, 96, 156
  method, 93, 94
CQF, 145, 146
Critical-band, 7
Cutoff, 38, 122, 153
CVSD, 13

D*PCM, 118–120, 122
D/A, 11, 14, 25, 28, 31

D_TO_A.BAT file, 169
Decimator, 147
Decorrelate, 153
De-emphasis, 87, 88, 95, 99, 102
Delta Modulation, 75
    adaptive, 78
    cvsd, 80
    linear, 76
Demultiplexed, 87
Density, 17, 32, 40, 46
DFT, 123
Differential, 67, 68, 70, 75, 107, 108, 115, 146, 155
Discrete-time, 85
Distortion, 11, 41, 46, 94, 99, 124, 132, 143, 145–147, 149, 150, 156
Dither, 153, 154
DM, 11, 75
DoD, 13
DOS, 25, 26, 28, 29
Downsample, 122
DPCM, 11, 12, 68, 70–72, 74, 75, 107, 108, 110, 116–120, 122
DPCM-AQB, 72
DPCM-AQF, 72
DSP, 1, 2, 9, 15, 86, 91
DSPLAB, 70–72, 120, 150, 153–157
    configuring, 25
    data range, 29
    destination directory, 18
    destination directory, 18
    directory, 20
    environment, 17
    exiting, 25, 28, 29
    memory usage, 27
    plotting, 31
    quit, 29
DSPLAB:destination disk, 18
Durbin, 92, 94

EEG, 4
EGA, 19
Elf, 19, 25, 28, 31

Encryption, 13
Epson
    24-pin, 27
    9-pin, 27
Error weighting, 129
ETSI, 14
Excitation, 5–7, 10, 86–90, 96, 97, 100, 102, 103, 105, 107, 122–124, 127–129, 131–139
    regular pulse, 135, 136
Excitation model, 99, 129
Exiting, 25
Expand, 18, 35, 38

Feed-forward, 72, 108–110, 115–117, 122
Feedback, 72, 108–110, 115–122, 128, 129, 154, 156
FFT, 39, 103
FILES, 20
Filter, 1, 5, 7, 9–11, 38, 39, 41, 67, 75, 86–96, 99–104, 109, 114, 115, 117–119, 122, 124, 128–132, 138, 139, 141–150, 153, 157
    error-weighting, 130
    unstable, 94, 99, 114
FIR, 5, 124, 131, 139, 145
Fixed quantizer, 41
Floating point, 30
Floating-point, 30
Formant, 5, 86, 94, 98, 99, 119, 130, 132
Fricatives, 5, 6
Full-band, 122, 123, 141, 147, 148
Full-rate, 14

Gain adaptation, 59
Glottis, 5
Goto, 35
GSM, 14

Half-band, 38, 143, 148
Hamming, 38, 92, 95

Harmonic, 101, 103, 111, 123, 124, 132
Highpass, 94, 143, 145
HLP directory, 20
HMM, 1
Homomorphic, 5
HPGL, 27
HP Laserjet, 27
HP Paintjet, 27

IBM Laser printers, 27
IBM Proprinter, 27
IBM Quietwriter, 27
IIR, 88, 118
Ill-conditioned, 94
Inaudible, 42, 142
Intelligibility, 106, 107
Interpolate, 75, 143
ITU, 13

JDC, 14

Kaiser, 38

Language, 3
Language model, 4
LAR, 103, 156, 157
LDM, 75
Least-squares, 91
LMS, 110
Lock, 26, 32, 33, 37, 158
Logarithm, 49
Logarithmic Quantizer, 49
Log-PCM, 154
Lowpass, 9, 39, 41, 67, 94, 122, 143, 145
LPC, 1, 5, 10, 12, 40, 85–90, 92–94, 96–106, 108–111, 115, 116, 121–125, 127–132, 134–140, 142, 143, 155–157
    residual-excited, 122
    spectral estimation, 103
LSP, 102, 157
LTP, 108, 111, 114–116, 120–122, 125, 132–137

MAC directory, 20
Markov, 1
Mid-riser, 43, 45, 46, 70
Mid-tread, 43, 45, 70
MNU directory, 20
Mouse, 14, 21, 22, 24–27, 29, 30, 35, 36
MPLPC, 12, 14, 127, 129, 134–137, 139, 140
MS-DOS, 17, 18
Multiband, 147, 157
Multipulse, *see* MPLPC

NEC Printers, 27
NFC, 116, 119, 120, 122, 129
Noise feedback, 119
Noise-Feedback, 116, 118
Nonadaptive, 67
Nonlinear, 47, 123, 129
Nonuniform, 7, 42, 43, 46, 146, 154
Nonuniform quantizer, 46
NxtLck, 35
Nyquist, 41, 75

Octave-band, 143, 146–150, 157
Open-loop, 125
Optimum Quantizer, 51, 52
Oversampled, 75

PARCOR, 92, 102, 155–157
Parseval, 103
Passband, 38
Paste, 30
PATH, 20
PCM, 1, 10–13, 42, 45–47, 70–72, 74, 108, 118, 150, 153, 154, 157, 158
PCM-AQF, 72
PCM-ASB, 154
Peachtree, 19, 25, 28, 31
Peak-limiter, 119, 120
Phase, 11, 23, 127, 143, 157
Phonemes, 4
Physiology, 5

Pitch, 5–7, 10, 11, 13, 86, 87, 97, 101, 105–109, 111, 115, 116, 119, 122–125, 127–129, 132, 133, 140, 156
    detector, 86, 97, 101, 105, 128
    excited, 85–87, 90, 97, 98, 100, 106
    high, 38
    modification, 105
    period, 86, 90, 95, 100–102, 105, 106, 114, 137
    prediction, 111, 119, 132
Plotting, 17, 30, 31, 36, 40
Poles, 94, 130
PopMenu, 35
Post-filter, 118
Postscript Printers, 28
Pre-distorted, 94
Pre-emphasis, 39, 86–89, 94–96, 99, 104, 105, 136, 137, 139
Pre-filter, 118
Predictor, 6, 7, 68–75, 87, 89–94, 96, 100, 102, 104, 107–111, 114–116, 120–122, 128, 133, 135, 137–139, 146, 150, 155–157
    long-term, 104, 108, 111, 115, 119, 120, 128, 129, 132, 133, 137
    short-term, 111, 128, 130, 132, 137
Predictor order, 94
Printers
    Canon BJ and LBP, 28
    Epson, 28
    HP Laserjet, 28
    HP Paintjet, 28
    HPGL Plotter, 28
    IBM Printers, 28
    NEC Printers, 28
    Postscript, 28
PrintGL, 28
    registration, 28
Printing, 25
PrvLck, 35

Pseudo-periodic, 100

QMF, 144–146
Quadrature, 144, 157
Quantize, 10, 41–47, 49, 68–72, 74, 75, 86, 87, 89, 102, 108, 110, 111, 115, 116, 118, 120, 121, 135, 142, 150, 154–157
Quantizer
    adaptive, 41, 53
    fixed, 41
    logarithmic, 49
    nonuniform, 46
    optimum, 51, 52
    uniform, 43

RAM, 18
Real-time, 129
Recognition, 1, 4
Recognizable, 106
Rectangular, 72, 73, 95, 133
Rectification, 123
Recursive, 5, 87, 88, 92, 93, 138
RELP, 122–124, 128
Residual, 7, 12, 71, 91, 92, 94, 100, 101, 103, 115, 116, 122–125, 127, 128, 135
Residual-excited LPC, 122
Robustness, 110
RPE-LTP, 14
RPLPC, 136, 137, 139

Sampled, 9, 41, 43, 75, 88, 101, 143, 148
Samples, 1, 29, 30, 33, 37, 41, 43, 46, 67, 68, 72, 88, 90, 96–98, 101, 102, 105, 108, 109, 111, 122, 133, 141, 148
Sampling rate, 1, 2, 29, 38, 42, 75, 88, 105, 124, 143
Satellite, 2, 10
SBC, 7, 142
SEGSNR, 43, 45, 70, 74, 75, 116, 120
SEV, 12, 129, 136, 139
Short-term, 67, 68, 107, 108

Signal-to-noise ratio, 42, 47
Singular-value, 138
Six-band, 147, 148
SNR, 14, 42, 43, 45–47, 70–72, 74, 75,
        110, 111, 115, 116, 118–120,
        122, 124, 136, 137, 139, 140,
        147, 149, 154–156
Software, 14, 15, 42, 101, 156, 157
Sound, 3
Sound Blaster, 19, 25, 28, 31
SPC directory, 20
Spectral estimation, 103
Spectrum, 7, 40, 96, 102, 103, 108,
        111, 116, 118, 119, 123, 124,
        129, 130, 142, 143, 146
  analysis, 86
  folding, 124
  shaping, 116, 130
  tilt, 95
  translation, 123
Speech communications, 2
Square-law, 123
Stationary, 11, 90, 107
Step size adaptation, 56
STFT, 1
STP, 108, 111, 114, 115, 120, 132
Stress, 4
Subband, 7, 13, 141–143, 146–150,
        157, 158
Subband coding, 143
Subband signals, 142
SVD, 138
Syllable, 4, 43
Syntactic, 4
Synthesis, 4, 75, 86, 88–90, 94, 100,
        102, 103, 106, 114, 122–124,
        128, 130–132, 138, 141, 147–
        150

Telephony, 2, 13
Texas Instruments 800 Printer, 28
TIA, 14
Time domain, 85, 91
Time frequency, 141, 148

TMS32010, 19, 25, 28, 31
Toeplitz, 92, 93
Toll-quality, 6, 11, 12, 71, 74, 85
Tree-structure, 143, 146, 157
  analysis-synthesis, 147
Turbulent, 5
Two-band, 13, 143, 144, 146, 147, 157
Two-band analysis-synthesis, 143

Uniform quantizer, 43
Unnormalized, 30
Unquantized, 10, 156
Upsample, 39, 148

Variance, 70, 72, 75, 125
VEV, 122
VGA, 19
VLSI, 1, 2, 86
Vocal tract
  time-varying, 90
Vocal Tract Model, 4
Vocal tract model, 88
Vocoder, 85
VSELP, 14

Whiten, 99, 104, 115, 117
Wide-band, 40
Window, 21, 23, 25–27, 45, 72, 73, 86,
        90–98, 100, 133, 135, 137–
        139, 154
  FIR design, 38
  length, 95, 97
  types, 95